IMAGES
of America

GOLDFIELD

Pictured are author Ted Faye (left) and Bryan Smalley. In August 2021, Faye spent nearly four hours with Bryan Smalley of Goldfield. Smalley owned Hidden Treasures Bookstore. It was more than a bookstore, though. It was a repository for the history of much of central Nevada, and that repository was Smalley himself. While Smalley was not that old, one would swear that he had been in Goldfield since its discovery in 1902. Smalley's tour of the place included his recalling events, people, and buildings that had been there as though he himself had met the people, been at the events, and visited the buildings. Within weeks of this photograph, Smalley passed away. It was a shock to the community and a tragedy for his wife and family. They have, however, chosen to keep the flame burning and are determined to honor his legacy by keeping this famed boomtown's history alive. They continue to keep the store open so that visitors both now and in the future can find their own historical nuggets at Smalley's Hidden Treasures. (Ted Faye.)

ON THE COVER: By the time this photograph was taken around 1906, Goldfield was booming. Prominent in the background is the Northern Saloon owned by famed boxing promoter Tex Rickard. (Bryan Smalley Collection.)

IMAGES
of America

GOLDFIELD

Ted Faye

ARCADIA
PUBLISHING

Published by Arcadia Publishing
Charleston, South Carolina

Printed in the United States of America

Library of Congress Control Number: 2023937801

For all general information, please contact Arcadia Publishing:
Telephone 843-853-2070
Fax 843-853-0044
E-mail sales@arcadiapublishing.com
For customer service and orders:
Toll-Free 1-888-313-2665

Visit us on the Internet at www.arcadiapublishing.com

*To all those who preserve, restore, renovate, and revive
the legacies of America's legendary boomtowns!*

CONTENTS

ACKNOWLEDGMENTS

The story of Goldfield, Nevada, is well documented by scholarly authors such as Drs. Elizabeth Raymond, Russell Elliott, Hugh Shamberger, and notably the late Dr. Sally Zanjani, whose father had come for the Nevada gold rush. These fine authors and others have done excellent work in telling Goldfield's story in words and images. Dr. Sally Zanjani is quoted liberally in the captions and narrative of this book.

The purpose of this book is to provide a look into life in Goldfield through recently discovered photo albums and archived images. Those who are familiar with Goldfield's story will see that this is not a comprehensive history of the town; there is much that is not told here. Rather, it is a collection of snapshots that hopefully are organized into an entertaining and informative story. Most of the photographs are in private collections, though there are some from museum and university archives as well. One significant collector who has been willing to share photographs is Telly Eliades (TEC) and his wife, Caroline. They are not just collectors, however, as they have purchased property in Goldfield and other former boomtowns and are working to preserve them for future generations. "Bullfrog" Bill Miller (WMC) specializes in the areas of Rhyolite and Bullfrog but discovered he also had some significant material relating to Goldfield and shared it here. The Borax archives at Death Valley National Park (DVNP) provided some rarely seen images. The Central Nevada Historical Society (CNHS) in Tonopah with its amazing collection and the help of Arlene Melton and her staff contributed to the book. The founder of the Central Nevada Historical Society and Museum, Allen Metscher, has a long and rich personal and family history in Goldfield, much of which he has graciously shared for various projects. Kenzie McPhie at the University of Wyoming, Frank Crampton Collection (FCC) provided some significant rarely seen Goldfield images. Others came from the Library of Congress (LOC) and the Society of California Pioneers (SCP). The digital collection at the University of Nevada, Las Vegas (UNLV) contributed some images, and the late Bryan Smalley's collection is noted as the Smalley Collection (SC). Grateful thanks and appreciation to the National Automobile Museum's (located in Reno, Nevada) staff and volunteers who spent a great deal of time and effort to identify many of the unusual vehicles seen in a variety of photographs in chapter four. The author's collection is noted as (TF). With all of these contributions, the process of how a boomtown "booms," the revelation of its daily life, significant events and people, and how it sadly comes to an end will take the reader through the rise and fall of America's last great gold rush town: Goldfield.

INTRODUCTION

For anyone visiting Goldfield today, the likely reaction could honestly be, "What in the world was the big deal about this place?" That question assumes, though, that the visitor knew it had been a big deal at one time. As of this writing, there is not a working gas station in town and only two motels and a couple bars, including Nevada's second oldest continuously operating bar, the Santa Fe Saloon, which serves some food. There are some food and drinks at the Mozart and some pretty good eats for breakfast and lunch at the Dinky Diner, and that is pretty much it. There is a general store that sometimes is not open at all. There is a lively radio station and some small businesses trying to make a go of it scattered around. Highway 95 cuts through the center of town, which connects Las Vegas in the south to Reno in the north, so a lot of traffic makes its way in and out of town. It is hard to say how many of the hundreds of people who pass through each day in those cars, trucks, motor homes, and on motorcycles know anything about the buildings they are passing or why the whole town is designated a national historic district.

The irony in all of this of course is that much of modern-day Las Vegas and Reno owe their existence and place in America's popular culture to this central Nevada boomtown. While Tonopah's 1900 silver strike, 25 miles to the north, certainly kicked off the 20th-century mining rush to Nevada, it was the gold discovery in the middle of the desert two years later that captured the imagination of the world. Not since the California Gold Rush of 1849, the Comstock Lode of 1859, or the Klondike Gold Rush of 1896 had America, or the world, seen anything like it. It was, in fact, the last great gold rush on the American frontier. And make no mistake about it, it *was* a frontier. And it *was* a gold rush.

Goldfield is in Esmeralda County, Nevada, which was established in 1861. Author Sally Zanjani in her book *Goldfield, The Last Gold Rush on the Western Frontier* quotes early historian Myron Angel describing the county as a "barren, unknown, waste," which, Zanjani says, it continued to be throughout the 19th century. In 1864, Nye County was carved out of Esmeralda, leaving it to around 7,000 square miles of desert so remote, harsh, and wild that most settlers happily left the land alone for the Native Americans to contend with.

The events that would lure tens of thousands of those who were not Native Americans to this vast wasteland began in 1900. It was the dawn of a new century, but Nevada was on the verge of losing its statehood. That option was being discussed in the halls of Congress as it seemed there was no good reason to maintain a state with so little to offer. All of that would change with Jim Butler's discovery of silver ore near some springs the Indians called Tonopah. The story goes that Butler was prospecting with his burros and picked up a rock to toss at one of the lazier ones, and lo and behold, it was full of silver. He told his wife, Belle, of the discovery but continued to hang out at their ranch. Belle thought Jim was procrastinating, encouraged him to get moving, came with him to the springs herself, staked her own claim (one of Nevada's richest), and soon the town of Tonopah sprang up. It attracted miners and prospectors to the region, and with Tonopah as a base camp, exploration began in the surrounding desert.

However, because this "wasteland" was the home of Native Americans, it is only fitting that the actual location of Butler's great silver strike and the upcoming gold claims to the south were first known to the Indians. While Jim Butler's story about his burros makes for good Western lore, it is more likely that Native Americans showed him the location of the silver. Some claim that Butler got the information from his Indian mistress or that he was taken to the spot by a Shoshone Indian prospector named Tom Fisherman. While none of the stories regarding Tonopah's silver discovery can be entirely proven, it is certain that Fisherman was responsible for finding a rich sampling of gold about 25 miles south of Tonopah in the location that would later become Goldfield.

According to Sally Zanjani, Fisherman was born in 1868 in an area known as Fish Lake Valley, about 50 miles southwest of present-day Goldfield. He had a wife named Minnie and was described as "intelligent and restless." He had a knack for prospecting, and for more than a decade went on prospecting trips by horseback, often taking his son with him. It was near a place called Rabbit Spring where Fisherman found the gold that caught the attention of Tonopah investors, along with two prospectors named William Marsh and Harry Stimler. Marsh was a rancher from Austin, Nevada, and Stimler, whose mother was Shoshone Indian, was born in Belmont, Nevada, and later came to work in the Tonopah mines looking for his big break.

Stimler sought out Jim Butler for a grubstake, where an investor provides support in exchange for a share of whatever the prospector discovered. "Grub" was the Old West term for food, and "stake" meant a share or portion of the mine's worth, which the investor would receive. According to author Carl Glasscock, the grubstake Stimler received from Butler consisted of "a dilapidated buckboard, a double team consisting of a horse and a mule, a bale of hay and grub for a week or so."

In December 1902, Harry Stimler and William Marsh went prospecting near Rabbit Spring, led by Tom Fisherman. It was a two-day trip; today, it takes about half an hour or so. It is possible that Fisherman and Stimler were related through Stimler's Indian mother, and thus, Fisherman was inclined to help Stimler and Marsh find the gold. Others suggest that Stimler beat Fisherman into telling him where the gold was. While, as Zanjani states, no one will ever know what happened that December day, it is certain that enough gold was found to forever change the face of the desert, Nevada, and the western frontier.

Fisherman had called his original claim Gran Pah, meaning "land of much water." *Pah* is a Shoshone word meaning "water" and is found in many Nevada town names, such as Tonopah, meaning "greasewood water," or Weepah, meaning "rain water." When Stimler and Marsh staked their claims near Columbia Mountain not far from Rabbit Spring, they kept Fisherman's Gran Pah name, making a play on words and stating that the mining district would be called the "grandpa" or granddaddy of all mining districts in the region.

Stimler and Marsh staked 19 claims, including one called Sandstorm because the discovery was made as great gusts of wind swirled sand around them. When rumors spread of Nevada businessmen investing in Stimler and Marsh's claims, a rush started for the district, but it was short-lived, as prospectors found nothing that lived up to the hype. There would be another false start for prospectors when another rush began some months later; however, the rush that lasted for several years began late in 1903.

Charlie Taylor and Alva Myers arrived and were given some unused claims by Stimler and Marsh. Taylor would go on to strike the rich Florence Mine, and Alva Myers would strike the fabulously rich Combination Mine, the first to ship gold ore. By October 1903, there was enough excitement in the Grandpa District that a group of 36 miners, prospectors, and other assorted interested folks got together to organize and create the beginnings of the town that Alva Myers dubbed Goldfield. Having "gold" in the town's name would certainly be a lure for investors, and unlike many hopeful discoveries throughout Nevada and the West, the quantity and richness of the gold in the district actually justified the name.

By 1905, less than two years from the 1903 organizational meeting, Goldfield had grown into a town of nearly 10,000 people. There were four banks and several mills; electricity was brought in from Bishop, California; and the recently completed Tonopah railroad significantly reduced the transportation time to the main line Carson & Colorado Railroad. When the Tonopah railroad was completed in July 1904, mule teams only had to haul items 25 miles from Goldfield to Tonopah to connect to the rail line, then the railroad would go the remaining 60 miles to the Carson & Colorado railhead at Sodaville. From there, shipments could go to San Francisco or points east via the Central Pacific. On September 12, 1905, a standard-gauge train line connected from Goldfield to Tonopah. It was now possible to take a train from Goldfield to just about anywhere in the country. Goldfield was on its way to becoming Nevada's largest and most popular city.

Carl Glasscock in his book *Gold in Them Hills* wrote, "The Deseret News of Salt Lake City on December 16, 1905, announced that a dozen Goldfield mines had already produced nearly seven million dollars. The Florence led the list with $1,848,000; the Combination was next with $1,800,000, and the Jumbo third with $1,000,000." These figures were based on net returns, according to Glasscock. One million dollars in 1905 equals more than $33 million today.

It was 1906 that became Goldfield's true boom year. Glasscock describes that year:

> Goldfield . . . was producing $130,000 [$4 million today] every twenty-four hours. The town had a population of fifteen thousand and everything in the nature of business and social activity that any mining town could ask or imagine. There was not a vacant store, office, house or room in the town. Buildings were under construction to house ten thousand more people. The increase in the market valuation of Goldfield securities for 1906 as registered on the mining stock exchange was one hundred million dollars [$3.3 billion today].

Glasscock interviewed the first woman to reside in Goldfield, Mrs. L.L. Patrick, who was the wife of the town's first successful promoter. She recalled,

> We had a wonderful social life. Outsiders, persons who have never lived in such a town as Goldfield probably would have been amazed to see our homes, our furnishings, our table service. The common belief, based upon fiction of the Bret Harte or Rex Beach type, is that mining camps are altogether tough and crude. That is not true . . . [we] had homes as well furnished, as well provided, as well organized and sophisticated as any New England, New York or southern city. To be sure, the tough and the crude were only a block down the street. They were available but unnecessary.

From 1904 to 1907, Goldfield experienced a building boom. Hugh Shamberger in his book on Goldfield writes, "One of the extraordinary features of the town of Goldfield was its development from a few tents and some dugouts and flimsy shacks in early 1904 to a small metropolis by the latter part of 1907, complete with many large stores, brick and frame office buildings, a large number of beautiful homes and a population of about 20,000." All of this was due to the fact that there was a group of fabulously rich gold mines.

Goldfield witnessed the evolution of transportation from horseback, burros, buckboards, stagecoaches, and mule teams to the railroad and automobile. Sally Zanjani notes, "as early as 1905, two 'automobile stage lines' carried passengers between Goldfield and Tonopah, more than halving the time consumed by the stage ride." Auto trips could cost up to $149 ($5,000 today) if travel began in Las Vegas and ended in Goldfield. Rich men and celebrities came to town in their autos—men like Charles Schwab, the Pittsburgh steel magnate who invested heavily in the mining boom; Death Valley Scotty, who claimed a secret gold mine in that most feared of deserts, Death Valley; and then there were the men who promoted and boxed in the "fight of the century."

When Tex Rickard came to Goldfield from the Klondike rush in Alaska, he had learned the art of promoting and staging fights from famed lawman Wyatt Earp. Both Earp and Rickard made their way to Tonopah and Goldfield, respectively, with Earp establishing the Northern Saloon in Tonopah and Rickard setting up his own Northern Saloon in Goldfield. Earp did not stay long in Tonopah, but his brother Virgil became Goldfield's sheriff and died in town of pneumonia on October 19, 1905.

In 1906, an event came to Goldfield that changed the world of sports forever and set the stage for boxing's reign both in New York and Las Vegas. After forming the Goldfield Athletic Club with some of the town's most influential people, Tex Rickard invited the lightweight champ, African American Joe Gans, to fight Oscar "Battling" Nelson, also known as the "Durable Dane," in a fight to the finish. Both men agreed and spent nearly a month in Goldfield prior to the fight. It was billed as a battle of the races, and the promotion was nonstop up until the day of the fight, September 3, 1906. The fight had reserved seats, was telegraphed blow by blow across the country, and was touted as "the fight of the century." Under a searing sun with temperatures over 100 degrees, the fight dragged on for 42 rounds until Gans was declared the winner on a foul. It had been agreed that no matter the outcome, Nelson would receive $20,000 ($660,000 today) and Gans $10,000 ($330,000 today) of the $30,000 purse.

The next year, 1907, brought the great financial Panic of 1907. Stocks crashed, banks closed, and the end of the last great gold rush was near. While Goldfield's mines continued to produce vast quantities of gold, the days of the rush and excitement were beginning to pass.

There were fires in 1905 and 1906, a flash flood in 1913, a major fire in 1923 that leveled 54 square blocks of the town, and a follow-up blaze in 1924 that destroyed much of what was left. Today, Goldfield is only a ghost of its former glory days, but there are dedicated individuals who seek to save and restore the structures that remain. As in Goldfield's early days, it is the people who believe and continue working to keep the town and its legacy alive. While tourism and some businesses are its primary reason for existence, there are those who believe it is important to give visitors a glimpse of the glory and excitement that was Goldfield.

One

DIGGINGS

There are certainly remnants of Goldfield's first settlements, but it takes a watchful and knowledgeable eye to spot them. The humble dugouts and tents on platforms are long gone, but some foundations and indents can be seen in the hillsides, evidence of Goldfield's early excitement.

It seems there were several "rushes" to Goldfield beginning in May 1903 when a rumor swirled that prospectors Harry Stimler and William Marsh had sold an interest in their claims to wealthy Nevada investor and financier George S. Nixon. While a virtual throng of prospectors swarmed the new gold region, it was not until later in the year when the Goldfield district was fully organized, the initial townsite laid out, and the first ore shipped in November that the true rush began.

Though there was no newspaper in the 1903 camp of Goldfield, Tonopah was only 25 miles away, and the Tonopah papers promoted the new strike. "Thar's gold in them thar hills," proclaimed the press. Two rushes happened in 1903, and the big one that "stuck" occurred in 1904. In the earlier rushes, prospectors came by the hundreds and were largely disappointed, as summed up eloquently years later by writer James F. O'Brien and quoted by author Hugh Shamberger in his book *Goldfield*: "The enthusiasts saw the billions of their crazy hopes dwindle to millions, the millions to thousands, and the thousands to the price of a meal."

But many of those who hung on to their claims and waited patiently were later rewarded with Goldfield's success. It was May 1903 when Alva Myers and R.C. Hart located the two Combination Claims and Charles Taylor located the Florence Mine. From October 20 to October 26, 1903, the Goldfield Mining District was organized, the townsite laid out, and the townsite company formed. In November 1903, the first shipment of gold ore went out from a claim called Combination No. 2.

But Goldfield still needed its own paper, for when it came to competing for investment money from people like Pittsburgh steel king Charles Schwab, the Tonopah papers clearly had their own interests at heart promoting Tonopah mines and investment opportunities first. On April 29, 1904, Jimmy O'Brien printed the *Goldfield News*, "All that's new and true of the greatest Gold Camp ever known." Goldfield had arrived.

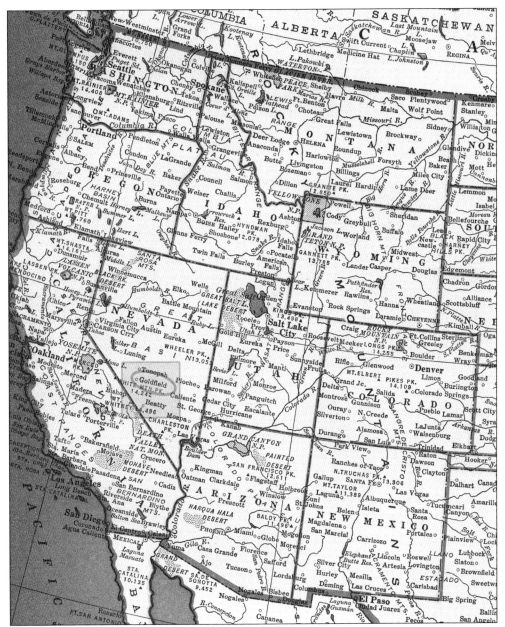

Goldfield is located where the Great Basin and the Mojave Desert meet. Two large areas known for their desolation and harsh environments come together in the Goldfield region. This map shows its location in the West as well as its proximity to Tonopah 25 miles to the north, Beatty 60 miles to the south, Death Valley approximately 80 miles southwest, Los Angeles 350 miles southwest, and Las Vegas 180 miles southeast. (TF.)

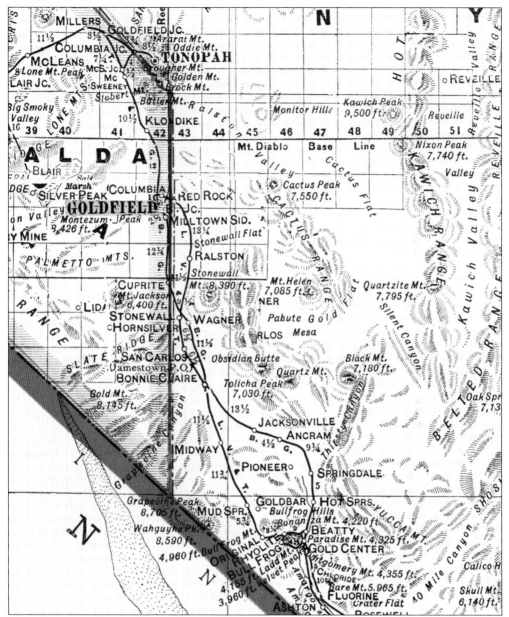

This detail of a c. 1910 map shows how much mining activity there was in the region. Place names and small towns fill the map where today there is only sagebrush. The solid black lines are the railroads to the region, of which there were three primary railways: the Las Vegas & Tonopah (LV&T), built by copper king William Andrews Clark; the Tonopah & Tidewater (T&T), built by borax king Francis Marion Smith; and the Bullfrog & Goldfield (B&G). Just south of Beatty, the solid line to the right is the T&T and the line to the left is the LV&T. At Beatty, the B&G (later merged with the T&T) turns to the right, and the LV&T heads off to the left. (UNLV.)

Sally Zanjani in her book on Goldfield describes the region as "those sand valleys and pinon-covered hills where few but the occasional prospector cared to venture, the land of terror commonly known as the 'black belt.' " After the great silver discovery of 1900 in Tonopah, prospectors began fanning out in the area looking for the next big strike. They were even willing to venture into regions they thought only the Indians could inhabit. (UNLV, Rinker.)

While there are no photographs of Shoshone Indian Tom Fisherman showing the first gold location of the "Grandpa" claim to Harry Stimler and William Marsh, these very early Goldfield prospectors could easily stand in for the trio. Though originally discovered in 1902, it would not be until the end of 1903 and early 1904 that the rush to Goldfield truly began. (SC.)

First picture taken in Goldfield, December, 1903. Harrie Taylor (discoverer). Lew Finnegan and John Y. McKame on the Jumbo, from which was taken $100,000 in 47 feet from surface, the strike that caused the first rush.

This photograph is captioned as the "First picture taken in Goldfield, December, 1903." However, so much of the caption is incorrect that whether or not it is the "first picture" is also in doubt. It was Charlie Taylor (Harrie, his brother, lived in the mining town of Bodie) who developed the Jumbo, though he did not discover it. It was Stimler and Marsh who laid claim to it and gave the undeveloped property to Charlie. What does seem to be accurate and is not disputed by any scholars is that $100,000 in gold was taken in 47 feet of digging. The first shipment of ore, however, was from the Combination Mine in November 1903. It shipped from the mine to the railhead at Candelaria by mule team, a distance of about 65 miles. (UNLV, Beckley Collection.)

The light-colored mounds in this c. 1903 photograph of Goldfield show the tailings from prospector diggings. From the looks of the photograph, it may seem fairly random as to where they would dig, but most prospectors could read the land. They could tell from surface quartz and type of soil where gold might be located. Most learned how to read the land from the Shoshone and Paiute Indians in the region. (SC.)

15

This late 1903 photograph shows snow on the ground and a variety of rugged shelters. Historian Russell Elliott writes, "In the first years in each camp's development, the isolation which forced the use of primitive mining methods caused similarly primitive living conditions. Consequently, the growing community was usually a hodgepodge of dugouts, tents, half-tent, half-frame houses, wood and gunny sack houses, bottle houses, barrel houses, stone houses, houses of mud and manure, or of any other material or combination of materials which could be found." (SC.)

"The distillate oil can became a recognized building material in these regions," writes Russell Elliott. "When filled with sand it could be laid like bricks; if opened flattened out, and nailed shingle fashion, it served as roofing material." This prospector at his rock shelter/dugout in 1903 has all he needs to find his fortune. One of the oil cans described by Elliott appears to be the top part of the chimney, with the flap of the can able to shut the flue as needed. (SCP.)

Wood was scarce in the desert camps so the resourceful builder used crates and stone and dug into the hillside to construct a home. The metal stovepipe peeks through the roof, which looks to be a combination of boards, stones, and dirt. There must be something of value inside as the door is clearly padlocked. (SC.)

Native Americans adapted to the invasion of the gold seekers, as seen in the combination of traditional and modern influences on this dwelling, including the hinged wooden door. The National Park Service, in a survey to create the Goldfield Historic District, reported that the area had been inhabited by Native Americans for thousands of years: "The earliest inhabitants are generally thought to have occupied the area at least 12,000 years before the present and the Western Shoshone still live in the region today." (LOC.)

Gold Field

NOV. 1903

When prospectors first arrived at the Goldfield district, they camped at Rabbit Springs near the mountains, as it was the closest available water but some distance from the diggings. When water was struck only 20 feet below the surface down on the flats, the tent camp relocated. According to Hugh Shamberger, "The location of this well was at what subsequently became the intersection of Main Street and Crook Avenue in Goldfield. About September 1, 1903, the small group of tents

at Rabbit Springs was moved down to the well site. The few miners camped around the new well felt that the tents should be arranged in some semblance of order; accordingly, early in September, what was to become Main Street in Goldfield was staked out." In this November 1903 photograph, Rabbit Springs is in the canyon at top center. (TEC.)

The beginnings of more permanent wooden structures began to sprout amidst the tents as lumber was brought in. Russell Elliott wrote, "The first permanent structures in [these boomtowns] generally came from older camps whose glory had departed. Such were the early frame buildings at Tonopah which had been hauled from Candelaria and Belmont. One of these, after serving as a

MAIN ST.
GOLDFIELD, NEV
JAN. 1904.

saloon at Tonopah, was moved to Goldfield to become that camp's first wooden structure." Hugh Shamberger wrote, "By 1904, the new town contained about 350 inhabitants, and a few frame buildings had replaced some of the tents. Over 15 business establishments were in evidence, with more springing up each day." (SCP.)

By mid-1904, there were five townsites in the district: Goldfield, October 1903; Milltown, November 1903; Columbia, February 1904; Diamondfield, February 1904; and Jumbotown, June 1904. This photograph is of Columbia in 1904, looking southeast. Within a year, Columbia had a newspaper, school, post office, saloons, and hotels. (UNLV, Rinker.)

HE GOLDFIELD NEWS

"All that's New and True of the Greatest Gold Camp Ever Known"

L. 1. No. 1.   GOLDFIELD, ESMERALDA COUNTY, NEVADA, FRIDAY, APRIL 29, 1904.   PRICE 10 CENTS.

The Greatest Gold Camp Ever Known--Its Story

WHAT TWO PROMINENT MINERS SAY OF GOLDFIELD.

[Newspaper article columns largely illegible]

No mining camp could survive without its own newspaper. James O'Brien opened the *Goldfield News* with an old printing press and published its first edition on Friday, April 29, 1904. Sally Zanjani writes that O'Brien printed it on "an old hand press welded together by an ingenious blacksmith from an assortment of scrap metal." The message was unapologetic and emblazoned in its banner: "All that's New and True of the Greatest Gold Camp Ever Known." Though it was not true in 1904, the phrase was nearly prophetic as Goldfield did become one of the richest gold mining towns in the entire West. In 1906, O'Brien sold his paper at a substantial profit. (LOC.)

Two

BOOMTOWN

The discovery of gold or silver in the ground was one thing. Developing that discovery into a sizeable town was quite another. Most western frontier mining towns never reached the size of Goldfield. And that is saying something because Goldfield was not near any resources that could help it grow as a town. It was, quite literally, in the middle of nowhere. When Tonopah was discovered, though remote, it was only 60 miles from the closest railhead. That may sound far, but ever since the 1870s when borax was discovered in central Nevada, mule teams had been hauling more than 150 miles to the railhead, and in the 1880s, teams hauled 165 miles from Death Valley to the nearest train.

Goldfield's growth was due to one thing: the continual discovery of ore rich with gold. The Combination Mine shipped the first gold ore in November 1903, then came the January Mine, the Jumbo, and the Florence. These four were the major producing mines until April 1906, when the Mohawk Mine was developed, and it was, in fact, the richest gold mine in the world. That boosted both Goldfield's population and building boom. Goldfield was actually built on fields of gold, and it certainly looked like it was here to stay.

Water and power were brought to town, railroads arrived, and multistory buildings were erected made of brick and stone. Goldfield came alive with true wealth surrounded by hopes and dreams. This chapter focuses primarily on the structures, the visual evidence of the phenomenal growth of an actual city in the midst of America's most remote desert in the earliest years of the 20th century. Ultimately, Goldfield had one of the finest hotels in the West and boasted theaters, banks, and a courthouse built to stand the test of time. There is no doubt that the little railroad town about 180 miles to the south that eventually grew into Las Vegas owes its existence to the mining camp that boomed into the city of Goldfield.

Supplies for building Goldfield were brought over the 25-mile route from Tonopah. Mule teams brought lumber, goods, and everything needed to establish the site. The lead wagon appears to be a 14-mule team driven by a teamster riding the left wheeler (wagon left), the mules closest to the

wagon. He is clearly holding the jerk line, which falls over his mule's right ear. This was a single line going to the lead left mule and was used to steer and control the team. A steady pull took the team to the left, and a series of jerks turned the team to the right. (SC.)

This mine, the Combination, shown in 1904, was struck in 1903 by the man who would later be called "The Father of Goldfield," Alva "Al" Myers. It is possible the man sitting in front of the person with his hands on his hips is a young Al Myers. Myers and his prospecting partner Bob Hart staked the claim despite Hart's protestations. "Stake, hell! This is no good!" exclaimed Hart, according to author Carl Glasscock. But Myers knew better, and the Combination became one of Goldfield's richest, making over $3 million ($97 million today) and paying out over $1 million ($32 million today) in dividends between 1904 and 1908. (UNLV.)

Prospector Charlie Taylor left Tonopah in May 1903 and headed down to Goldfield to try his luck with Stimler and Marsh's Grandpa District. Stimler and Marsh, according to author Sally Zanjani, "offered him the Jumbo, one of their lapsed claims. With that casual piece of generosity, Charlie Taylor received a piece of ground destined to become one of the premier mines of the district and to eclipse the Marsh and Stimler claims. Taylor also located the Florence of future fame." From 1904 to 1908, the Jumbo would produce more than $1.7 million ($55 million today). (UNLV, Rinker.)

Charlie Taylor also found the Florence Mine, seen here with miners standing next to a primitive headframe and suspended ore bucket in 1904. The triangle-shaped frame over another part of the mine is called a "whim" and the mule is hitched up to a line that runs through the pulley wheel seen at the bottom of the front leg of the whim. The line goes to the top pulley and is then connected to the chain, which carries the ore bucket up and down the mine shaft. (UNLV, Rinker.)

This view is below the Florence Mine where the ore brought up out of the ground is sacked and ready to be shipped. It will be taken by mule team to the mill where it will be crushed and the gold extracted from it, then likely vaulted and prepared for sale. (UNLV, Rinker.)

In this wide shot are the two mines that produced the first shipments of gold. This c. 1905 photograph shows both the Combination Mine and the January Mine. The Combination Mine (1) was the first mine to make a shipment in November 1903. The January claim (3) struck in January 1904 and

This image from the same perspective as above shows the fully developed Combination Mill completed in 1906. The development of mills near the mines was significant for Goldfield as ore

was the second mine to ship out gold. Here, the foundations (2) are being laid for the Combination Mill, and in the background is the town of Goldfield (4). (UNLV, Rinker.)

could be processed on-site and raw ore did not have to be shipped for miles to a remote mill to have the gold extracted from the rock. (UNLV, Rinker.)

This 1904 view of Goldfield shows how the town was primarily tents with a few wooden structures scattered throughout. Laid out in relationship to Columbia Mountain at left, the population continued to grow significantly as more gold was found. The *Tonopah Bonanza* newspaper wrote on February 13, 1904, "The rush to Goldfield continues with unabated vigor and the excitement over the showing so far is increasing daily. . . . Many tent houses and more substantial buildings are being erected and the music of the hammer and saw is heard from early morning until after dark." (UNLV, Rinker.)

Like many other Nevada mining camps, water was rarely found near the discovery sites. While Rabbit Springs was not close to the major Goldfield discoveries, it was soon found that shallow wells could be dug on the flats, providing water for the young camp. But the development of more mines and the influx of more people meant much more water was needed. Water companies such as the Goldfield Water, Mining, and Milling Company; the Goldfield Water and Transit Company; the Esmeralda Water Company; and the Montezuma Water Company were formed. There were also six 3- to 5-inch pipelines built with a total length of 47 miles, the longest being from the springs near Lida (north of Death Valley) at a distance of 36 miles. This gushing artesian well is from one of the early drillings in 1904. (UNLV, Rinker.)

"Goldfield is to have an electric light and power plant . . . Whereby the offensive odors and inconvenience attendant upon the lamp and candle are to give way to the soft glow of the ever-ready incandescent, the citizens have every reason to be congratulated." Published in the *Tonopah Bonanza* on June 4, 1904, this article revealed how quickly the camp was growing, stating that more than 1,000 people were coming to the area per month. While a steam-generated plant provided power beginning around January 1905, it would be the Nevada Power, Mining, and Milling Company that would transport power all the way from Bishop Creek in California over nearly 100 miles of transmission lines to the growing town of Goldfield. This transformer building was part of the power infrastructure continuing to be improved as late as 1908. (UNLV, Rinker.)

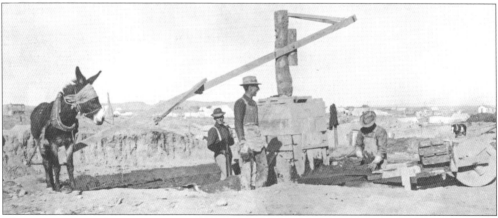

The year 1905 saw the first major building boom. According to the December 29, 1905, issue of the *Goldfield News*, "Three-quarters of a million dollars were spent on buildings . . . and to a great extent frame and adobe houses have taken the place of primitive tent-houses, and improvements in residences will be especially numerous as the tents and tent-houses which fringe the more densely settled confines of the city will rapidly disappear, and in their places substantial frame and adobe cottages will be built." Here, a burro is used to manufacture adobe bricks. All of the stone and bricks for buildings were produced locally. (SC.)

The next 15 images show the evolution of Main Street with Columbia Mountain distinctly visible in the background. In late 1904, Goldfield was beginning to look like a typical western town. A wooden boardwalk was constructed in an attempt to avoid the dust and muck of the town's dirt streets. Note the facade front of the Hardware Store at the far end of the street. (SC.)

The Hardware Store, seen here, housed a variety of businesses, including a drugstore. The Snug Saloon was next door, and the Goldfield Undertaking and Mercantile Co. Inc. offered both mortuary services and tent supplies. The tent building at the end of the block advertised beds for 75¢ a night. (SC.)

Main Street is shown here in either late summer or early fall in 1904, with a mule team coming down the middle of the street. Both sides of the street were being developed, and substantial wood buildings included a telegraph and telephone office and a jeweler. At the far end of the street on the right is a building with a peaked roof, the Colorado, which provided rooms for rent. (SC.)

The camera has moved down the street and the peaked building, the Colorado, is visible at right, advertising rooms. The backwards sign off the porch of the Colorado advertises "Private Dining Rooms Upstairs. Now Open." In every image of Main Street, it is interesting to note the consistent flurry of activity. (SC.)

In October 1904, this automobile showed up, creating considerable interest and curiosity from the men bystanders. It appears that the driver (in uniform at center) is standing by as one of the men takes the wheel to try it out. This early-20th-century gold rush was unique, as automobiles (or "flivvers" as they were called) drove alongside prospectors with their burros and mule or horse-

drawn wagons and stagecoaches. The peaked roof of the Colorado can be seen in the distance on the right. On the left, the white balcony of Ole Elliott's saloon can be seen. The front of the saloon housed Goldfield's first bank, established July 23, 1904—the State Bank and Trust Company. (UNLV, Rinker.)

In December 1904, the State Bank and Trust moved into one of the first stone structures in town. The building can clearly be seen at left. Across from the bank is the peaked roof of the Colorado, and the Palace Saloon is visible to the right. The man driving the wagon at left is dressed in typical western worker's attire, including the light-colored suspenders. (SC.)

It appears that there is snow in the foreground. At nearly 6,000 feet, Goldfield's winters can be brutal. For as hot as its summers can get, its winters can be cold, windy, and snowy. The building at left with the flag is the newly built and unpainted Esmeralda Hotel. Note the barber pole in front of the hotel. This image was likely taken very late in 1904, possibly December, as the stone State Bank and Trust can be seen down the road on the left. (SC.)

One of the busiest street scenes captured of Goldfield, this image illustrates a boomtown in its bustling growth phase, with wagons bringing in goods and supplies and men working and inspecting new buildings going up. Tents are long gone, and substantial wood, brick, and stone structures are underway. This photograph was likely taken very late in 1904. (SC.)

A 14-mule team hauls a freight wagon through town. Note the driver of the team is riding the left mule closest to the wagons. He guides the entire team with one line leading to the front left mule. That line is called a jerk line; the driver pulls on, or jerks, the line in order to steer the team. Today, Highway 95 runs between the Palace Saloon and the building across the street with the stacked firewood. This intersection, the corner of Main and Crook Streets, was destined to become the focal point of town, as one saloon would eventually be on each corner. The State Bank and Trust resides in one of the first stone structures in town. The bank would eventually become the Hermitage. Across the street from the bank would be the Northern saloon, and across from the Palace would be the Mohawk. (UNLV, Rinker.)

This image of Goldfield shows a newly painted Esmeralda Hotel and the Miners Union Hall under construction just up the street. Goldfield was growing at an unprecedented rate and would soon

be the largest city in Nevada. (FCC.)

eet Scene.
Goldfield.

In 1905, Tex Rickard came to town and opened the Northern saloon. It was in the Colorado building and soon became one of the most popular bars in town. Rickard arrived in Goldfield fresh

from the Alaskan Klondike rush. There, he had learned much about running a bar and promoting fights from famed lawman Wyatt Earp. (SC.)

In 1906, just as Goldfield was experiencing a major building boom, the richest discovery of all was made. The Mohawk became Goldfield's richest mine, producing over $6 million in gold in its first

GENERAL VIEW OF THE
MOHAWK MINE, GOLDFIELD, NEVADA.
THE GREATEST GOLD MINE KNOWN.
FIRST SIXTY DAYS' OPERATION $6,000,000!! PRODUCTION.

COPYRIGHT NO. 4200
P. E. LARSON PHOTOS
GOLDFIELD, ESMERALDA COUNTY, NEVADA, U.S.A. 1906

1. COMBINATION MILL
2. JANUARY MINE.
3. GOLDFIELD.
4. FRANCES MOHAWK LEASE.
5. TRUETT LEASE.
6. HAYES & MONNETTE LEASE.
7. MACKENZIE LEASE.
8. SHLET & ISH LEASE.
9. CURTIS BROS. LEASE.
10. ODDIE LEASE.
11. EL CANEY MINE
12. COLUMBIA MOUNTAIN.
13. SILVER PICK MINE
14. NEVADA GOLDFIELD REDUCTION CO.
15. R. R. STATION.
16. RED TOP MINE.
17. KIA KEAD MILL.
18. KEWANA MINE.
19. LAGUNA MINE. 20. WELLS FARGO EXPRESS CO. LOADING ONE CAR, FOR SHIPMENT $1,000,000
(ONE MILLION DOLLARS) GOLD ORE.

two months. This overview shows the intensity of Goldfield's mining activity and its proximity to the town itself. Note the key and its corresponding numbers. (UNLV, Rinker.)

At the intersection of Hall and Main Streets, this view is looking in the opposite direction of the previous series of Main Street images (facing west) with the Malpais (Malapai) Mesa in the background. Note the Hardware Store, still in business three years later when this image was taken in 1907. The Nevada-California Power Company was on the second floor of the State Bank and Trust until the company moved late in 1907 to its own building. The stone building on the left midway down the street is the Nixon Block, constructed by Sen. George Nixon. It housed the John S. Cook Bank, one of Nevada's most successful early banking firms. (TEC.)

Goldfield had a sizeable red-light district, so-called due to the red-colored lights prostitutes would place in their windows letting potential customers know they were available. "ladies of the night," "working women," "sporting women," and "soiled doves" were all names for prostitutes. While most think of prostitutes as only providing sexual services, they often performed wifely duties for single men as well. Darning socks, mending clothing, cooking, and sometimes just offering companionship were other services provided. In the middle of the red-light district, Victor Ajax opened an elite restaurant and dance hall called the Ajax Parisienne. (TEC.)

Exploration Mercantile Co. was an early mining supply store built in 1904. It was the first large building in Goldfield and, by 1906, also housed the *Goldfield Gossip* newspaper. It was an essential type of store in a mining town, providing prospectors and miners everything they would need for their work and exploration. (SC.)

At the far end of Main Street toward Columbia Mountain, the San Carlos Hotel was built in 1906. Not much is known about the San Carlos, although it was one of Goldfield's early hotels. The photograph has a few curious elements to it and raises some questions as to what the items are hanging on top of the hotel. Are they bedsheets and blankets, washed and now hung out to dry? Is the ladder there for workers or for someone to get in the hotel window? The answers will likely never be known, but the image provides insight into early Goldfield. (UNLV, Rinker.)

The first Goldfield Hotel is seen here in 1906. It was one of Goldfield's finest hotels, constructed entirely of wood. It opened November 1, 1905. It had 53 rooms, all equipped with hot and cold water and electric lights. Private baths were attached to some of the rooms. (UNLV, Rinker.)

This magnificent shot of the first Goldfield Hotel features an electric light over the front entrance. A variety of well-dressed Goldfield folks stand in front of the hotel in 1906. (SCP.)

This rather haunting photograph outside the first Goldfield Hotel is from early 1906, apparently just after Christmas 1905. A man can be seen walking behind the tree in front with his dog closely following. A man watches from the window where Victorian lamps and furnishings can also be seen. The hotel was destroyed by fire on November 17, 1906. (TEC.)

One of the town's earliest stone structures was the elegant Nixon Block. It was built by Nevada senator George Nixon, who served in the US Senate from March 4, 1905, until he died on June 5, 1912. He partnered with powerful businessman and entrepreneur George Wingfield in the John S. Cook Bank Company. The Cook Bank would be housed in this building, which cost $80,000 to construct on a plot of land costing $10,000. It was a three-story building with a footprint of 50 feet by 100 feet at the northeast corner of Main Street and Ramsey Avenue. It was completed on September 25, 1905. (TEC.)

In 1907, Goldfield's premiere hotel opened for business. The Hotel Casey was one of the finest hotels between San Francisco and Denver. This advertisement features the hack services available between the hotel and all incoming and outgoing trains. The Hotel Casey was true luxury in the desert. (TEC.)

The Hotel Casey was a magnificent building at the corner of Main Street and Miners Avenue. It was three stories high and 75 feet by 100 feet in size. It housed 36 two-room suites with a bath in every room. It even had an elevator. In 1908, George Wingfield and his Bonanza Hotel Company bought the Hotel Casey and shut it down in order to drive business to their new and grand Goldfield Hotel. (SC.)

Gold was literally everywhere in the early days. One newspaperman wrote, "I saw a man pull up a bit of sagebrush and pan the dirt from the roots, getting several good colors." Here, as the new Goldfield Hotel is under construction in 1907, miners sit on the tailings from their mine. The windlass (the crank used to haul up buckets of ore and lower supplies into the shaft) can be seen behind the miners. This operation was just behind the Goldfield Hotel. The old Goldfield Hotel, shown in previous images, was located on this spot but burned down in 1906. (TEC.)

Two large buildings are going up in 1907, the Goldfield Hotel at far right and the Montezuma Club/News Building with the white stone and arched doorway in the foreground. On December 1, 1906, the *Goldfield News* ran an article stating that the owners planned to erect "a $200,000 brick hotel building, and its construction will begin as soon as the plans, which are now in the hands of a Reno architect, are completed." In November 1907, the *Goldfield Review* stated, "The new Goldfield Hotel, which is rapidly nearing completion, will . . . contain 175 guest rooms . . . and will have cost upwards of $300,000." (UNLV, Rinker.)

Newe Bldg & Goldfield Hotel.

The arched doorway in the foreground building led to the Montezuma Club. The club was established in 1904, meeting in the home of one of Goldfield's earliest resident businessmen. The purpose of the club was "to have a place where visitors could be entertained; at the same time, business matters could be discussed with some degree of privacy," as Hugh Shamberger states in his book *Goldfield*. By 1906, club membership had grown to 600. For a while, the club met above the Palace Saloon. Then in 1907, backed by one of the mining barons, the new building was funded. The quid pro quo, however, was that the mining baron would be president of the club. The *Goldfield News* shared the building with the club. (UNLV, Rinker.)

In 1907, the Goldfield Hotel is nearing completion. Scaffolding and construction supplies can be seen outside the building. Behind the man standing is the Montezuma Club/News Building. (TEC.)

The Goldfield Hotel was dedicated on January 15, 1908. It was not quite complete, as only the ground and second floor could accommodate guests. This early advertisement promotes all of the wonderful amenities available at the hotel. What was once sagebrush and lonely desert now looked like a major street in any big city in America at the turn of the 20th century. (TEC.)

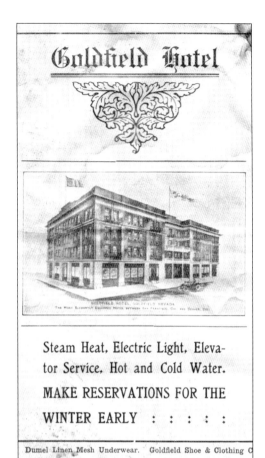

The Goldfield was elegant for any location. It would have done well in San Francisco, Denver, or Los Angeles. The dining room is seen here immediately after the hotel opened in 1908. Author Sally Zanjani describes the dining room displaying "floors of tiny black and white mosaic tiles, potted palms, and heavy white linen tablecloths." It was the largest room in the building and was as wide as the structure with plate glass windows overlooking Crook Street (today's Highway 95). The restaurant was dubbed "The Grill." (UNLV, Rinker.)

GOLDFIELD HOTEL LOBBY

The lobby of the Goldfield Hotel is described by author Sally Zanjani: "Its mahogany trimmed lobby was resplendent with gilded columns, black leather upholstery, and globular chandeliers." There was even an elevator that ran 300 feet per minute, a switchboard, and a public telephone booth. (UNLV, Rinker.)

The Goldfield Hotel was the most impressive building in town. It was four stories with 154 rooms, all with running water, though guests did have to share a claw-foot bathtub and toilet. The hotel opened over a three-day celebration that was funded by a $10,000 loan, but it was paid back with all of the partying, drinking, and likely gambling that took place at the hotel over those three days. It still stands today, though sadly many of the interior furnishings have been inexplicably and irresponsibly removed or destroyed in recent years. (UNLV, Rinker.)

Located at the corner of Columbia Street and Ramsey Avenue, the Nixon-Wingfield Building (with the window awnings) was a three-story stone structure built in 1907. Owned by Nevada senator George Nixon and businessman George Wingfield, it was also home to one of Goldfield's largest mining operations, Goldfield Consolidated Mines. Next door was the Ish-Curtis Building, constructed by Frank and Marvin Ish and Uri Curtis. They established the Nevada Registration and Trust Company where mining companies could register their stocks so there would not be overissues of stock to protect the investors. Both buildings still stand today. (TEC.)

City leaders wanted culture to be a part of Goldfield life, and the Hippodrome Theatre was built for such entertainment. The large wooden structure went up in just 28 days in May and June 1907. It was enormous for such a town. It sat 1,200 people, and the stage was 34 feet wide by 35 feet deep. Three 50-man shifts worked on building it in record time, but in 1920, it was sold for $1,000, and it took only seven men to dismantle it in 27 days. (TEC.)

On June 18, 1907, Goldfield's castle-like courthouse was dedicated; however, it did not actually open until 1908. The wooden, slatted pylons in front of the building were to protect the small trees that were planted. Part of the reason for the courthouse was that the Esmeralda County seat had been moved from Hawthorne, a small town to the north, to Goldfield. Ron James, in his book *Temples of Justice*, notes that the conflicts between miners and mine owners, including the use of federal troops, had made the working situation so tense that James calls this "one of the harshest courthouse facades built in Nevada . . . with all the coldness of a remote Western federal penitentiary." (SC.)

The Goldfield Bank and Trust opened in December 1904 prior to the completion of the building pictured above and was located in a corrugated iron structure next to the new building under construction. Unfortunately, the Goldfield Bank and Trust went out of business on May 23, 1905, less than a year after it opened. Bank auditors found only $21.05 left of the $80,000 worth of deposits made. The president of the bank and the cashier made their way to San Francisco, where they were later arrested. Sometime after the Bullfrog & Goldfield Railroad and the Tonopah & Tidewater Railroad merged in 1908, the Tonopah & Tidewater opened offices in the former would-be bank. The Tonopah & Tidewater was built by borax king Francis Marion Smith and connected his borax operations and far-flung mining towns to the Santa Fe at Ludlow, California. (DVNP.)

This is Columbia Street with the Ish-Curtis Building (far left), the Goldfield Hotel on the left, and the News Building. The image is prior to 1919 as the John S. Cook Bank has not moved into the Ish-Curtis Building as of yet. Rather, there appears to be a bar called Kays Atlanta Bar and another restaurant or club with its wooden sign posted above the door beside it. (DVNP.)

The most exciting intersection in town was the meeting of Crook (Highway 95 today) and Main Streets. Here, a drilling contest draws a crowd of thousands. At far left is the Hermitage Bar. At far right, the marquee for Tex Rickard's Northern saloon is seen. Across the street from the Hermitage is the Mohawk, and across the street from the Northern is the Palace. This was at the height of the town's growth. Note the multitude of electric wires and phone lines nearly obscuring the action. (SC.)

This panorama from 1909 was shot from the top of the four-story Nye Building. Prominent in the foreground is the Nixon Block (center). Just behind the far left corner of the Nixon Block can be seen the pointed cupola of the Nevada Hotel. It was touted as one of Goldfield's finest, with 28 suites with bath and 100 sleeping rooms with the best possible plumbing and heating. The hotel was scheduled to open in 1908, but construction shortcuts and shoddy workmanship led to it being condemned. While a large and imposing structure in Goldfield's skyline, it never opened for business. Up the street to the right of the Nixon Block the two-story stone building with awnings

directly behind the block is the M&M (Monian and Mitchell) Building, which housed the offices of the Tonopah & Goldfield Railroad. After the M&M was destroyed by the 1923 fire, the Elks Club was built on the site. Catty-corner from the M&M is another building with awnings, which is the Nixon-Wingfield Building housing the Goldfield Consolidated Mines offices, and to its right is the Ish-Curtis Building, home of the Registration Trust and 10 years later, the John S. Cook bank. Behind those two can be seen the three-story Goldfield High School with the flag flying prominently. (LOC.)

This page from the *Goldfield Daily News* on Saturday, January 9, 1909, details the growth of the town from mining claims to a city with money and nightlife. From the dugouts and primitive shelters of early days to the "many well-built and handsomely furnished houses in Goldfield that afford all the comforts of 'civilization' having steam and hot water heat, electricity for light and cooking and every accessory of modern houses," Goldfield's daily life was beginning to resemble life in any big city. (TF.)

Three

DESERT MARVEL

"Goldfield is the most metropolitan and cosmopolitan camp of all . . . a marvel of variety, color and life," exclaimed poet-prospector Clarence Eddy. Author Sally Zanjani uses the quote to explore how diverse Goldfield's population was and illustrate that people of various classes interacted with each other. The line between rich and poor as well as job distinctions seemed to disappear in Goldfield. The *Goldfield News* reported that "tenderfoot, gambler, clergyman, legitimate promoter, mining shark, sure thing man, panderer, courtesan, each have their place in the passing show."

On a daily basis, life in Goldfield was frenetic, exciting, anticipatory, and exhilarating. And it could also consist of the ordinary. People went to work, they went sightseeing, they shopped, took care of business, traveled, hung out with friends, ate, danced, played sports, went to school, gambled, drank, and celebrated holidays. Though primarily populated by white men, there were women, children, and minorities. The African American community was "quite a settlement," according to one Goldfield resident. Native Americans who found their homeland invaded yet again tried to adapt as best they could to the ways of the white man.

As author Carl Glasscock notes, "Considering the number, variety and instability of its residents, the great sums of easy money put into circulation by its high-graders [miners who hid gold in their clothing when they left the mine, keeping it for themselves], the amount of liquor available and the extent of the gambling that went on, Goldfield was a mining camp remarkably free from bloodshed. Gun-play was exuberant rather than bloody."

Though there were technically five towns in the mining district (Goldfield, Milltown, Columbia, Diamondfield, and Jumbotown), Goldfield was the leading town and continued to be so. By 1907, there were 15,000 to 18,000 residents. It is hard to know exactly how many people were there because Goldfield's boom happened after the census of 1900, and by the 1910 census, the town was in severe decline. But while it bloomed, it was indeed a desert marvel.

Goldfielders knew they owed their existence to the Indians. While they may not have recognized Tom Fisherman as the original discoverer, they could not ignore the fact that Harry Stimler was half Shoshone. Though Stimler was celebrated, it did not do much for the Native Americans who had to survive in an economy they did not understand. Captioned as "Our Indian Neighbors," this image was in a photo album belonging to a young lady of Goldfield. Her name is unknown, but she captioned the images with dates and first names of people she knew. These Indians may have been neighbors, but they wanted no part of the camera. (TEC.)

These two men in a mine represent the camaraderie, friendship, and like-minded intensity of purpose of Goldfielders. The man on the right is Cleveland Rinker, who came for the Goldfield gold rush from Indiana. Rinker never worked in the mines, but in the offices of the mining companies. Mine tours were popular, and this photograph emphasizes the importance of the mines to everyone in town, as they produced some of the world's richest ore. Both Rinker and his friend hold a candle. In the early days of mining, candlelight was the primary means of lighting a mine shaft or tunnel. This photograph was taken around 1905. (UNLV, Rinker.)

This businessman with his rolltop desk represents the everyday work that might be conducted in an office. The calendar reads January 1905. There is a panoramic photograph of Goldfield on top of the desk, and an electric lightbulb provides light. This office could be in any modern city at the turn of the 20th century. (SC.)

The men in this drugstore are dressed in their business finest—three piece suits on the gentlemen in the store and a bow tie and suit on the clerk behind the counter. It is unclear what he is holding, though it appears the man on the right may be making the purchase. The calendar appears to read February 1905. (SC.)

Not only did Goldfield have electric lights, but it also had telephones. These operators are working what is clearly a busy switchboard. Keeping all of the wires straight was a daunting task, and listening in for gossip was likely irresistible. In fact, newspaper editors complained that breaking news phoned in by a resident often was disconnected, resulting in the gossip mill cranking out the story before the paper could print it. (TEC.)

Goldfielders must have loved going to the soda fountain. The flavors can clearly be seen as orange, coffee, chocolate, strawberry, lemon, what appears to be sarsaparilla, and vanilla. The young boy at right is likely learning from the other two more seasoned "baristas." (SC.)

The ride from Tonopah to Goldfield was 28 miles but often went very slow as the road could be crowded. Sometimes wild mustangs were recruited for the team as freight animals were in such demand. Fred Corkill, superintendent of Pacific Coast Borax's mining operations, noted that the stage from Tonopah to Goldfield was a "dusty ride of six hours from Tonopah on a rig called a stage, three in a seat built for two." This image was taken around 1905, prior to the building of the train. (SC.)

Everyone knew that a mining town's biggest enemy was fire. Wooden structures, canvas, and all manner of flammable material in a mining camp made it pure kindling for any nearby spark. The Goldfield Fire Department was first organized in 1905 and was all volunteer. However, by 1906, some of the firefighters must have been paid, as a newspaper article reported they walked off the job for lack of payment. This 1908 image shows several members of the firehouse with the engine in front of the Montezuma Club at the News Building at Columbia Street and Crook Street (Highway 95). (SC.)

Goldfielders needed wood for their stoves to cook and to stay warm in winter. George Cramer's wood yard, seen here in 1907, likely provided a fine livelihood. Cramer in his big button shirt and suspenders and his worker in overalls must have put in many hard days of work chopping wood for the townsfolk. (TEC.)

While some Goldfielders were still using horses, mules, and burros, the more affluent were using autos. These vehicles required service stations, so a new type of business sprang up. Here, a couple is standing in front of the authorized Ford dealer in town; however, that did not prevent the Ford dealership from hanging a Buick sign in the window, making them a one-stop automotive shop for Goldfield. A delivery or freight truck is behind the couple. (TEC.)

This image of several storefronts shows the diversity of the businesses. The Goldfield Assay Office is at far left. The Goldfield Produce & Commission Co. is shown at center with "Fish Specialties"—in the middle of this remote desert. Its delivery vehicle sits out front laden with wooden crates. The Columbian offered furnished rooms. (TEC.)

Here, the Exploration Mercantile Co. is seen fully decked out after being in business several years. Everything a miner, prospector, or adventurer would need could be found in this store. There are standing pine trees outside the store along with men who represent a wide variety of classes and occupation, each dressed differently. (SC.)

This image from 1907 shows Mary's Dance Hall advertising the arrival of the Sells Floto Circus. The Lew Dockstader Minstrels were also coming to town, likely playing at Mary's establishment. Goldfield's dance halls were an important part of the social life of Goldfielders. Author Sally Zanjani quotes one writer describing the dance halls with "a crowded dance floor, uproarious with laughter, shrill chatter, the scraping of fiddles and the tinkling of pianos plunking out waltzes and two steps and girls, the principal attraction in woman-starved Goldfield . . . the Girls wore low-necked dresses with skirts above the knees and painted themselves like totem poles." (LOC.)

This 1908 image is the interior of the Bank Saloon, which was in the Nixon Block at Ramsay and Main Streets. On the back, it notes that Charles Evans purchased the Bank Saloon from Ole Elliott and Kid Highly for $25,000. Ole Elliott was one of the founding fathers of Goldfield, having his own saloon on Main Street just down from the Esmeralda Hotel. That $25,000 was no small sum, as today it would be worth $3.2 million. (TEC.)

These ladies have prepared a sumptuous banquet at the Miners Union Hall. While this image is somewhat later, in 1911, it does show the preparations and types of food for a banquet. There is a good amount of breads, cakes, fruits, and what appears to be pie. The dishes for the main entrée appear to be upside down in anticipation of the meal being served. (TEC.)

Members of the black community in Goldfield are seen at a cakewalk dance contest labeled "Ham's Grand Cake Walk." According to an article by Victoria Linchong, the cakewalk originated on plantations in the 1840s, stated with slaves satirizing the formal European dances of their masters: "Throwing their shoulders back and tilting their heads haughtily, they linked arms and promenaded side by side down a chalk-line. Soon, it became a contest. Whoever could move the most gracefully and pivot with the greatest ease at the end of the chalk-line would win a cake." This was the origin of the phrase "takes the cake" and "cakewalk," meaning something easy. However, the cakewalk only appeared simple as the dancers moved with great skill. It is likely whites participated in blackface alongside African Americans. Cakewalks were a part of minstrel shows from the 1870s until the early 20th century when vaudeville began replacing minstrel shows as entertainment. (SCP.)

The Goldfield Hotel is decorated for Christmas around 1908, and guests are dressed in their finest. Note that the tree goes all the way to the ceiling and is nailed to a wooden stand. It is strung with electric lights. Sally Zanjani notes that "the Christmas season, in addition to its spiritual significance, brought a heightened round of parties and dances. Huge bunches of red peppers decorated store windows, children received handsome gifts from the Elks Club, and in 1907 the Mohawk Saloon hosted a free Christmas dinner for 1,400." (TEC.)

In Goldfield's early days, there is no doubt it resembled the old Wild West. A dirt street with horseback riders and horse-drawn carriages is evident in this rare image. There are also poles for electricity and the new phone service, which indicate a more modern Wild West. An architectural business is getting set up in a newly painted storefront at right. There is an energy and excitement

reflective of the general optimistic mood and tone of early Goldfield. It should be noted that many who came for this rush were very conscious of the old West, and many tried to relive those days or play out their own Wild West fantasies during this last great rush. (FCC.)

All holidays paled in comparison to the Fourth of July. Parades, parties, contests, sporting events, patriotic speeches, and every type of excess could be found during the Independence Day festivities. Here, an early Fourth of July parade (c. 1906) passes one of the town's earliest hotels, the Esmeralda. Notable is the participation of the town's ladies: three are on horseback while other women stand on the balcony fully bedecked, including a parasol to protect from the sun. At far left, a man holds a banner for the local miner's union. (FCC.)

In what is only tersely captioned as a "Mule Fight," apparently two mules are down and their condition is unclear. Note that the mule at center is clearly concerned about what has just happened. The street seems clogged with mule teams. Mules, a cross between a male donkey (jack) and a female horse (mare), are estimated to be 30–40 percent smarter than either of their parents. They are pound for pound stronger than horses, can go longer distances, and, because of their donkey genes, are more suited to desert conditions. It is unclear what started this fight or how it concluded. More than likely, the man in the overalls to the right is one of the teamsters (drivers). (SC.)

This 1908 photograph of Goldfield High School shows the magnificent two-story building (and the basement) with students and teachers out front. Announcements for the construction of the school were made in 1906, but it did not open until January 1908. It had 12 rooms and officially began with 125 students enrolled. A number of kids are clearly clowning around (some things never change), and there are clearly younger students that would not yet be in high school. But in later years, the school housed grammar school students. The last classes were held in May 1953 when the school was closed for good. It still stands today and is being lovingly restored. (UNLV, Rinker.)

Sports were a part of Goldfield life. Baseball and football were both played on a field in an open area later used for "The Fight of the Century" boxing arena. Simply captioned "around the end" by the woman who put a 1907–1908 Goldfield photo album together, this image shows an active scene from an early football game in Goldfield. (TEC.)

This was a typical home in Goldfield in 1907. John and Myrta Fenwick are seen in their doorway. John Fenwick was uncle to Cleveland Earle Rinker, who was pictured in a mine with his friend on page 60. Rinker was from Indiana and met a man there who was investing in Goldfield. He headed west and stayed here with his uncle John, who operated a store in Goldfield and knew everyone in town. He helped Rinker get accustomed to daily life in this desert boomtown. Rinker wrote that life here was quite different from Indiana, from the high price of food and water to the 24-hour gambling and drinking halls to the periodic street shootings. (UNLV, Rinker.)

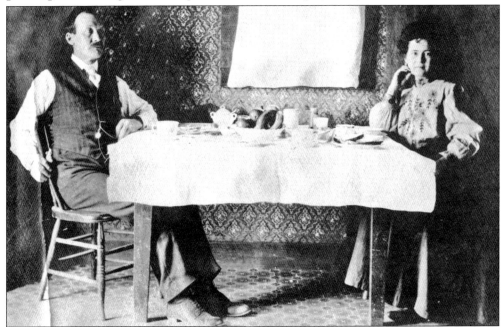

Inside the Fenwick home, John and Myrta pose for the camera. It could be any middle-class home in America at the time. The table is nicely set, the wall is papered, and the floor has linoleum, although the window curtain seems a bit primitive. John and Myrta Fenwick, however, are well dressed for the camera. (UNLV, Rinker.)

Behind their modest home, the Fenwicks pose with their cat while possibly John or Myrta's mother gives the dog a treat. Behind them is one of Goldfield's many mines. The cactus-like tree in the foreground is a Joshua tree, common in the high deserts of Nevada and California and unique to the Mojave Desert. (UNLV, Rinker.)

Labeled "Our Trip to the Park, 1909" in the Goldfield photo album whose owner's name is unknown, this image shows a hike to a vantage point overlooking the thriving metropolis of Goldfield. There is an additional caption for the photo reading, " 'Oh!' you 'Kid!' " In other images in the album, the men are referred to as "Kid," such as "Kid Louie" or "Kid Wallie," so it may simply have been a term of endearment. The man at center seems to be a little flirtatious and "handsy" with the woman on the right (who seems to be enjoying it all), while the woman at left, created the album, looks remarkably uncomfortable. (TEC.)

Here, the uncomfortable lady in the previous photograph seems to have lightened up with "Louie." The caption "The way Louie works" seems to refer to Louie's charm, judging by her smile. The view is west on Ramsay Street near the intersection of Fifth Street. The structure at right is the Presbyterian church. It was established in 1905, originally meeting in a tent, and pastored by Reverend Byers, who had moved from Oregon. After raising $10,000, the congregation built its new church and opened its doors on Easter Sunday in 1906. Remarkably, the wood structure survived Goldfield's fires but was dismantled in 1925 and moved elsewhere. As of the writing of Hugh Shamberger's book *Goldfield*, the church's new location remained a mystery. (TEC.)

Simply captioned as "Don L. and Jimmie," these two men appear to be going to or from the library, school, or church, judging from the armful of books. In the background can be seen the Presbyterian church to the right and the high school between the boys. Likely a photo from around 1909, this is farther down Ramsay Street than the previous image, possibly between Euclid and Fifth Streets. (TEC.)

This image from the photo album appears to be on Main Street. It is captioned, "Ethel–Alice, "Where art thou"–Roeina–Corena–Ruby." The woman who created the album is likely second from left. It is somewhat difficult to align the five names with the four people. One interpretation might be that Ethel is on the left and "Alice, where art thou" may indicate that their friend Alice is missing. Perhaps the mystery lady who created the album is Roeina (next to Ethel), and the other two are Corena and Ruby. Another possibility is that Alice was the album creator, and Roeina is the one missing. (TEC.)

Labeled as "Eating Candy, Jim, Walter, Jimmie, Wilk, Tom, Candy," the photo album creator continues to have a good time labeling her photographs. It is unclear who took the photographs. These men appear to be sitting on a beam above a foundation being constructed for a new building. This appears to be near the northwest corner of Ramsay Street and Columbia Street and is likely the foundation for the Monian and Mitchell Building (M&M Building), which was undergoing construction in 1907 when this photograph was taken. The steeple of the Presbyterian church can be seen in the background. (TEC.)

Labeled as "Our Trip to the Mines," the unidentified photo album creator must have been accompanying the men since she includes herself in the description. The mine they are visiting is "The Combination Mine," she writes, one of Goldfield's first mines. Mine tours were common from the earliest days of mining in America. People were curious as to where the minerals came from and how they were discovered in the rock. In the very early days of places like Cripple Creek, Colorado, some mine operators even gave very small "samples" to visitors, though it is not clear if Goldfield mine operators followed such practices. Seen here are some of the men from an earlier image. From left to right appear to be Wilk, Jim, Candy, and Jimmie. (TEC.)

Simply labeled with a big question mark above the photograph, this is a humorous image of one of the album creator's Goldfield pals. It appears to be Walter, or "Kid Wallie." He took off his hat for this photograph; it is lying on the ground next to him. A man wearing only a barrel was an early humorous reference to someone who was flat broke. Usually, the man was depicted with no clothes under the barrel. Clearly, this is the modest version. (TEC.)

The photo album does not miss Goldfield's landmarks, and one of the most important was the Las Vegas & Tonopah (LV&T) Railroad station. It was owned by Montana senator William Andrews Clark, who had built the line from the tiny connecting town of Las Vegas to the boomtown of Rhyolite, arriving in Goldfield by 1907. Now visitors could get to Goldfield by rail from San Francisco or Los Angeles. Here, in 1909, the station can be seen covered in snow. A passenger train has arrived at left, and it appears that there are freight cars in the distance to the right. By 1918, all of the towns on the LV&T were gone, and the line had no reason to exist. The rails were torn up and the metal used for the World War I effort. Sadly, only ruins of the station survive, but there are hints of its former elegance. (TEC.)

GOLDFIELD CON. MILL AT NIGHT.

ME

The photo album labels this image "100 Stamp Mill, Goldfield Consolidated Mines, Nixon-Wingfield." It dramatically illustrates Goldfield's reason for existence. In November 1906, George Nixon and George Wingfield created the Goldfield Consolidated Mines Company. It was a conglomeration that made them some of the richest mining men in the country. The giant mill ran 24 hours a day, seven days a week. The sound could be heard for miles around as 100 stamps smashed thousands of pounds of rock to get out the gold. It was a constant thunder that Goldfielders tolerated, for as long as those stamps crushed rock, their livelihood was assured. (TEC.)

The most direct reference as to who compiled all of these photographs into the album is this image labeled "me." While she can be seen in a number of other photographs, she never really refers to herself. The best that can be determined is that she is either Roeina or Alice, and everyone is thankful for the personal glimpse she has given into this remarkable desert marvel. (TEC.)

Four

CARS AND STARS

Flivvers, chug-chug machines, and desert yachts were all early names for the automobile. Along with electricity and the telephone, it was one of the primary innovations that distinguished this Wild West from the Wild West of Wyatt Earp and the romanticized West of Buffalo Bill. In Goldfield, wagons pulled by mules drove right alongside Fords, Ramblers, and a host of other not-so-well-known automobile brands.

The *Goldfield Daily Tribune* wrote on July 28, 1907, that "no automobiles used anywhere in the world get such hard and racking service as those in use in Nevada. It costs an automobile owner here between $1,000 and $2,000 a year easily for repairs." The paper continued, "A race for the location of mining claims between two ambitious operators who have thousands at stake may oblige the chauffeur to make a night run over unknown roads into a trackless country where any kind of an accident may happen to him . . . The credit for the development of Nevada . . . is due to the chauffeurs who guide the machines into all nooks and corners of the state."

By 1907, according to author Hugh Shamberger, most of the important people in Goldfield had an automobile. They had chauffeurs who were trained in changing tires and making repairs as well as driving over the rough desert roads. A Goldfield resident named Charles Chrisman even built his own auto, and the 1908 Great Race from Paris to New York went right through the heart of town.

Not only were there cars, there were stars! Goldfield also had its share of local celebrities like the original discoverers Billy Marsh and Harry Stimler, though for some reason Stimler was the more popular of the two. There were the Earp brothers, one of whom, Virgil, stayed on as sheriff for a while. Walter Edward Scott, known as Death Valley Scotty, was a regular in town before and after his trips to Death Valley. Well-known British novelist Elinor Glyn visited Goldfield, and actor Nat Goodwin performed at the Hippodrome. Tex Rickard built a nice brick house on Crook Street and became a famed boxing promoter, creating New York's Madison Square Garden as a mecca of boxing. And then there was Jim Casey, who went to Goldfield and started a delivery service that grew into the company known as United Parcel Service, or UPS. Not bad for a remote desert town that a few years earlier had been home only to sage, cactus, and coyotes.

Author Sally Zanjani writes, "After the first automobile to cross the desert . . . arrived at Tonopah from Sodaville in July 1903, the new invention became the rage. Establishments such as the Palace Blacksmith Shop began to advertise 'automobile repairing' as well as 'horseshoeing.'" Here, a 1904 Rambler is delivered in Goldfield via what must be one of the earliest car carriers. (SC.)

"The cheapest automobile in the world. Everybody should have one," read the advertisement for the Waltham Orient Buckboard. This 1904 model in Goldfield was manufactured in Waltham, Massachusetts, by the Waltham Manufacturing Company. According to Sotheby's, the Orient featured "a remarkably simple steel chassis with no meaningful suspension, a natural wood-finish body and fenders, and wire wheels, which were powered along the road by an air-cooled, single-cylinder engine. . . . It was the very embodiment of a horseless carriage." (SC.)

Pictured here is a 1905 Thomas Flyer. Erwin R. Thomas was no Henry Ford, but the car that he created and that bears his name is one of the most memorable of the early era of automobiles. According to Sotheby's, "The first Thomas car, the single-cylinder . . . debuted in 1903. For 1904 a new 'Flyer' model featured a three-cylinder 24-horsepower engine . . . In 1905, Thomas joined Franklin as one of the first American manufacturers to build a six cylinder." Sally Zanjani writes that a "new malady" called "Chauffeur's blindness" required drivers to wear goggles to protect from the dust, wind, and brilliant sun of the desert. (SC.)

According to Hyman Limited, the "Pope-Toledo grew out of the International Bicycle Co., one of Albert Augustus Pope's businesses. From 1904, the company offered steam, and later gas-powered cars. The gas versions proved quite successful in motorsport, with a Pope-Toledo coming in 3rd in the highly competitive and popular Vanderbilt Cup in 1904 and winning the America's first-ever 24 hour endurance race in 1905. Pope-Toledo cars grew swiftly in size and price in the coming years, culminating in the 50 horsepower limousine of 1907. This prestigious and beautiful machine sold for a robust $6,000 and was among the finest automobiles offered to wealthy American buyers." One of the fellows pictured here, who appear to have drunk their last beer (or some beverage), must have been wealthy indeed to afford such a car, if in fact one of them owned it. (SC.)

Two women are pictured in the drivers' seats. On the left is a Waltham Orient Buckboard. The vehicle to the right appears to be an early Pope-Toledo (likely 1903). The Motor-Car website states, "The 1903 model was an open two-seater with four wheels and front engine. It was powered by a

series end-cylinder engine with a removable cylinder head. The engine power was transmitted via a three-speed gearbox and one chain to both rear wheels. The car had a mainly wooden frame, which was combined with a steel subframe that carried the main mechanical components." (SC.)

On the left is a 1905–1906 Thomas, and on the right is a Winton Quad, likely a 1904 model. An advertisement for the Winton exclaims, "Uhheard-Of-Excellence . . . a car luxuriously comfortable, thoroughly substantial, safe and swift, and you describe the Winton Quad." According to www. secondchancegarage. com, Scottish immigrant Alexander Winton in 1891 opened a bicycle company, then five years later built his first single-cylinder automobile. In 1898, Winton sold 22 vehicles, including what is recognized as the first American-made, gasoline-powered, mass-produced car. When a young Henry Ford applied to be a mechanic at Winton's shop in 1899, Winton turned him away as he was not sufficiently impressed with Ford's skills. Here, the people appear to be in front of Carl D. Drossel's store, which advertised lots for sale in Sen. George Nixon's town as well as nearby Diamondfield. (SC.)

This early Pope Toledo touring car (c. 1905) is on display as an "Automobile for Hire." It appears that the hood has been removed to show off the engine. Judging from the legs visible under the car, there are five men in some kind of discussion. The chauffeur's head can be seen just behind the steering wheel, with hat and goggles. It is likely the crew is about to head out on a desert adventure. By 1905, there was regular car service between Goldfield and Tonopah 25 miles to the north. (SC.)

In 1906, one of the town's founders, Ole Elliott, along with some others, invested in local inventor Charles Chrisman's Chrisman Desert Flyer. Few photographs exist of the vehicle dubbed the "Monarch of the Desert," and the two that are known appear to show Charles Chrisman (possibly pictured here) with at least two women in the car. A 1907 *Goldfield Gossip* article reported that the Chrisman Desert Flyer "made the run from Goldfield to Rhyolite, eighty miles, in two hours and thirty minutes, taking the road 'as she lay!' " There were plans to build a Chrisman Flyer plant in Los Angeles, but the car was only ever produced in Goldfield. (SC.)

This image of a man, a dog, and his Cadillac, found in the personal photo album of Arthur Otis Eppler, has no caption. Eppler was both a singer and photographer. The credit reads "Allen Photo Co. Goldfield, Nevada." Allen was Eppler's stage name, used for his photo studios in Goldfield and San Francisco. He was described as a "wandering minstrel of the west's gold camps." In a 1930 *San Francisco News* article, a successful Eppler reflected on his days in Goldfield: "Those friends of mine still following the camps are most of them broke . . . but great fellows, the early day miners and mining camp men. They never forget a friend." (SCP.)

AUTO - RACING - ON - AUTO-L[

This 1908 photograph, taken on what is described as "auto-lake," shows what must be the beginning of an auto race with bystanders and viewers to the right and the cars coming toward the camera

NOV-22-1908-

- ALLEN-PHOTO-CO
GOLDFIELD-NEV-

on the left. The photograph is by photographer and musician Arthur Otis Eppler. The location is likely Alkali Lake northwest of Goldfield. (SCP.)

Between Tonopah and Goldfield, there were two auto stage lines as early as 1905. Here, a Winton Quad carries travelers to Goldfield. It cut the stagecoach time in half. There was a line that opened north from Las Vegas, making the 124-mile trip to Rhyolite for $40 and then an additional $25 for the 70-mile stretch from Rhyolite to Goldfield. Only the affluent could afford such prices, but in one year there were close to 9,000 passengers on just one of the Tonopah-Goldfield routes. (SC.)

Just 15 miles outside of Goldfield, this Winton Quad had a breakdown. The driver, wearing his goggles, bends over to assess the damage. Breakdowns were not that common, but there were accidents, even though the average speed was just 19 miles per hour. The first was in February 1905, when an auto stage turned too sharply and overturned. No one was seriously injured. Sally Zanjani writes that in 1907, the city had its first pedestrian hit by a car, and the first auto theft was reported. The town also had 40 private cars valued at $250,000. The local news lamented, "The way these big lurching bounding automobiles went 'tearing through the town;' and their riotous, ripping roistering honking became such a problem . . . that demand developed for a speeding ordinance . . . unknown in the days when local safety ordinances centered upon the need to hitch horses." (SC.)

In 1908, Goldfield became part of the Great Race from New York to Paris. On February 12, 1908, the race began in New York's Times Square. Four nations participated: Germany, France, and Italy with one car each, and the United States with three, including one Thomas Flyer. The American Thomas Flyer took a 500-mile lead, but 75 miles north of Tonopah (100 miles from Goldfield), its transmission cracked. One of the team members found a "flea bitten horse," rode to Tonopah, and got help. Some men drove from Tonopah to assess the damage, drove back to Tonopah, got the parts, drove back out to the Thomas Flyer, repaired the car, then drove back to Tonopah! On March 21, 1908, the Thomas Flyer arrived in Goldfield. (UNLV.)

Driving down Ramsay Street, the Thomas Flyer attracts a throng of Goldfielders. The Thomas Flyer team was treated to a banquet in town but soon left for Death Valley, arriving in San Francisco just three days later on March 24, 1908. The Thomas Flyer had traveled 3,836 miles in 41 days after leaving Times Square. On July 30, 1908, after traveling 169 days, 13,341 by land and 8,659 by sea for a total of 22,000 miles, the Thomas Flyer arrived in Paris. The team won the race. The car was shipped back to the United States and eventually became a part of the collection of the National Automobile Museum in Reno, where it can be seen today. (TEC.)

One Goldfielder wrote, "Scotty rides in an otto that makes more noise than a dozen ottos ought to!" Wherever Walter Edward Scott went, he made a splash. "Death Valley Scotty" was known as the "Monte Cristo of Death Valley" and a King Midas of the desert, but above all, he was known for his secret Death Valley gold mine. No one ever saw his mine, but as one old-timer expressed, "Everyone saw Scotty!" Here, as the caption states, is "Scotty's return from Death Valley." Everyone was anxious to hear what the man with the secret gold mine had found. Fans were mostly disappointed when Scotty would pull tricks like ordering drinks at the bar, then giving the bartender a $100 bill. No one had that much change in those days, and the bartender would begrudgingly "comp" the drink. Scotty is seen here sitting in the back seat with his legs crossed and wearing his signature hat. (SCP.)

There were always prospectors coming into and leaving town. Goldfield was home base to buy supplies, get a burro, and maybe find someone who would support their prospecting efforts in exchange for what was found, a grubstaker. They kept the prospector supplied with grub (food) in exchange for a stake (a percentage) of the gold that was found. Here, Cy Johnson is seen returning from a prospecting trip in Death Valley. Surrounded by humble burros, Johnson stands in contrast to the flash and flare of Scotty's would-be prospector antics. (UNLV.)

One of the celebrities who came to town was British novelist Elinor Glyn. While her visit included promoting local mining stocks with actor Nat Goodwin and boxing promoter Tex Rickard, she also observed the men of Goldfield. Sally Zanjani quotes her: " 'The types thrilled me,' said Glyn. 'Infinitely better bred-looking than any I have seen elsewhere in America, and some extraordinarily good looking . . . they might all have been princes and dukes by their manners . . . There among them were the most complete specimens of what we call breeding in England, that indescribable balance of limb . . . and their throat coming from the unfettered blue flannel shirts, with the points of the collars sticking up and no tie, showed proportions of young gods!' " In this photograph, it appears that Arthur O. Eppler, the erstwhile musician and photographer, is standing to her left. (SCP.)

Diamondfield Triangle Mining Co.

Capitalized
1,000,000 Shares

Par Value
ONE DOLLAR

Fully Paid and Non-Assessable

JACK DAVIS
President

GEO. WINGFIELD
Vice-President and Treasurer

WILLIS SEARS
Secretary

JACK DAVIS
Locator and Founder of Diamondfield Mining District, Nevada

Treasury Stock
400,000 Shares

50,000 Shares
will be sold at 15 cents
per share

Treasury stock, however,
is subject to advance
without notice

Direct your correspondence
to

Davis & Sears
Diamondfield
Nevada

The Diamondfield Triangle Mining Company includes the following property: Daisy Triangle, Lulu, Great Bend and Daisy Fraction, Elsie, Wednesday Fraction, and joins Goldfield Daisy Syndicate, Jumbo Extension, Goldfield Belmont, Goldfield Tonopah Mining Co., Red Butte Group and the Palace Claim.

The Diamondfield Townsite

also offers great opportunity for investment and profit. It is situated four miles east of Gold-field, and is surrounded by such mines as the Quartzite, the Black Butte, the Vernal No. 2, the Jumbo Extension, the Goldfield Diamond Co., the Great Bend, the Goldfield Daisy, etc.

DAVIS & SEARS

Original Locators and
Mining Experts

DIAMONDFIELD, NEVADA

SEND FOR FREE ILLUSTRATED BOOKLET ON MINES

One of Goldfield's most colorful characters who was present at the first meeting to establish the town in 1903 was Jackson Lee Davis, commonly known as Diamondfield Jack. In February 1904, two towns were established on the border of Goldfield within the same week, Columbia and Diamondfield. Its founder, also called Jack Davis, was the man who staked the rich claims, laying the foundation for the town. As Hugh Shamberger writes, "He was a living legend, having narrowly escaped a legal hanging in Idaho before going to Goldfield. His gunslinging activities around Goldfield added to his legendary fame." Sally Zanjani provides a portrait: "A man of medium height with a dark mustache 'grown long enough to chew on when angry,' Davis wore black broadcloth and a black sateen shirt and sported a 'mean-looking bowie knife, exhibiting it at frequent intervals as a slicer for his chewing tobacco . . . Davis [would] stride into a restaurant for supper, seat himself at a table facing the door, and ostentatiously place his guns beside his plate.' " (UNLV.)

One of Diamondfield's early mines is shown here in 1905. From 1904 to 1908, the Diamondfield mines produced approximately $500,000, more than $16 million today. (UNLV.)

Diamondfield was about four miles northeast of Goldfield's business district. It was able to support several businesses on its own, and there were a number of houses for mine workers and operators. A post office was set up in November 1904 but was discontinued May 30, 1908. From 1907 to 1908, there were about 150 people living in Diamondfield with two restaurants, four saloons, two rooming houses, and a grocery store. (UNLV.)

On October 1, 1909, the American Mining Congress held its 12th annual session in Goldfield. Here, in front of the spectacular Goldfield Hotel, the men gather for a photograph with a woman on the balcony seemingly overseeing the entire affair. There were many speakers at the event, and mining policy was discussed and set out in great detail. Two notable speakers were part of the meeting: *Los Angeles Times* mining editor Sidney Norman, who was on the advisory board for the organization, and Charles S. Sprague of Goldfield. It was interesting that Norman was on the board, as he had spent much time chasing Death Valley Scotty around the desert trying to see his mysterious gold mine. Sprague is of interest because he managed the strong vaults in the Ish-Curtis Building. Of course, George Wingfield was a director for the group as well. At some point during the proceedings—either before or after the 10:00 a.m. morning session, during the lunch break

from noon to 2:00, or in the break between the afternoon session between 5:00 and 8:00 when the last session began—this group photograph was taken. Also, note the cars. From left to right are a Pierce Great Arrow, Franklin, Thomas Flyer, another Pierce Great Arrow, and another Franklin. During the congress, Charles Sprague gave a rousing opening speech interrupted by thunderous applause several times as he outlined the history of the region: "As the rush overflowed at Tonopah the cry was 'on to the south' and at the foot of Columbia Mountain a pick was struck whose echo was heard around the world, and it was the signal for the revival of the greatest interest in mining and speculation the world ever knew. Within these few years—it was six years this month, when the first discoveries were made—within these few years, this little patch of ground has given to the world not less than thirty-five million dollars of the yellow metal!" (UNLV.)

James Emmett Casey (right), was born March 29, 1888, in Candelaria, Nevada, not far from Goldfield. In 1897, his family moved to Seattle and young Casey earned a living on the streets making deliveries for stores and businesses. Casey's father died in 1902 and he had to support his family. In 1903, he founded the City Messenger Service, delivering phone messages. In 1905, he sold the business as news was spreading of a new gold rush in Nevada. Casey joined the rush. Though he failed at mining, he and a partner delivered messages, making $50 a month. When Casey's partner was shot to death by one of the town's most disreputable characters, Casey went back to Seattle. Here, he is seen with business partner Claude Ryan in the Seattle office of their American Messenger Company, which eventually became United Parcel Service. (UPS archives.)

Wyatt Earp (left) came to Goldfield fresh from the Klondike gold rush. Earp had run saloons there and promoted and refereed boxing matches. When he arrived in Tonopah, he opened a saloon with business partner Al Martin called the Northern. He may also have gone to Goldfield and worked in Tex Rickard's Northern Saloon as pit boss for a while. Wyatt's brother Virgil (right) arrived in camp around 1904 and worked as a guard in one of the saloons, later taking on the role of sheriff. Goldfielders were unimpressed with his connection to the shootout at the O.K. Corral, as it had not yet risen to mythic status in popular culture. Virgil died of pneumonia in Goldfield in 1905. He was 62. Wyatt visited his brother while he was sick but left Goldfield for Los Angeles, where he lived another 24 years, dying in 1929 at 80. (Both, LOC.)

Actors Nat Goodwin and Edna Goodrich perform a scene from *The Genius* in 1907. This very scene would have played out at Goldfield's Hippodrome Theatre, as *The Genius* starring Nat Goodwin was one of the theater's offerings. Goodwin had a unique relationship with Goldfield as he was not just an itinerant actor coming through town but became actively involved in mining promotions with the notorious George Graham Rice and boxing promoter Tex Rickard. Goodwin came from a mining family and had become the most famous American comedian of the era. Edna Goodrich became part of Goodwin's theatrical troupe, and in 1908, two years after *The Genius* debuted on Broadway, Goodwin and Goodrich were married. However, Goodwin apparently got involved with another actress. In 1912, a nasty court battle ensued between the two, and a judge granted a divorce, allowing Edna to use her maiden name and barring Goodwin from ever marrying again in the state of New York. (Wikimedia Commons.)

This 1936 photograph shows Death Valley Scotty (Walter Scott) and Major Smith, who is described in the caption on the back as "the first negro to come to Goldfield, Nevada in the boom days when Scotty made his reputation as a desert Midas. [Scott and Smith] met recently for the first time in 32 years. They are shown discussing the colorful days they knew so well." (TEC.)

The last two celebrities pictured on this page are the "discoverers" of Goldfield. Seen here is Billy Marsh, who went out with Tom Fisherman and Harry Stimler on the north ridge of Columbia Mountain, found gold, and located three claims, the Sandstorm, Kruger, and May Queen, naming the new district "Grandpa." Marsh took his earnings from mining, married his childhood sweetheart, and bought a ranch. He had ups and downs but eventually settled in Tonopah, becoming a respected local politician and serving as a county commissioner, assemblyman, and state senator. He died in 1938 at the age of 62. The people of Tonopah paid him respect by lowering the flags and closing businesses in his honor. (CNHS.)

Harry Stimler's life stands in contrast to Marsh. While Marsh gave up the mining life, Stimler doubled down on it. But he became a promoter rather than a miner. He continued his relationship with Tom Fisherman, the Indian who actually discovered Goldfield and who brought even more discoveries to Stimler. Stimler got close to other big strikes but nothing again like Goldfield. He established a brokerage business that failed, and ultimately moved to Tecopa, California, southeast of Death Valley. Sally Zanjani describes what happened there: "On a midsummer morning in 1931, he was sitting in his car talking to his secretary in front of the Hall general merchandise store in Tecopa when Franklin Hall suddenly erupted from the store, shouted 'Now I've got you' and fatally shot Stimler, afterward committing suicide. The cause was apparently a trivial dispute over a borrowed rock crusher, which triggered a killing rage in the mentally unstable storekeeper." As for Tom Fisherman, he lived for a while from some of the money he made helping Harry Stimler. In January 1923, however, Fisherman had been drinking in a Tonopah saloon and, on his way home, somehow tripped and fell 78 feet to his death in Jim Butler's Silver Top Mine. (CNHS.)

Five

THE LONGEST FIGHT

If there is one event in Goldfield's short history that had some of the most far-reaching implications and consequences, it is "The Fight of the Century," as it was billed at the time. It was promoted as a "battle of the races." A black fighter and a white fighter would duke it out to the finish to become the lightweight champion of the world, and it would happen in Goldfield, Nevada.

Joe Gans, nicknamed "The Old Master," would fight Oscar "Battling" Nelson, known as "The Durable Dane," for the championship on Labor Day, September 3, 1906. The sponsor of the event was the newly formed Goldfield Athletic Club, comprised of businessmen, stock promoters, and George "Tex" Rickard, who had recently come to Goldfield with Wyatt Earp from the Alaskan Klondike rush. Earp had taught Rickard about promotion in a saloon where they worked together in Dawson. Rickard, along with others, believed a prize fight would promote Goldfield mining stocks and make the town a household name across the nation. Sure, it was becoming known as a gold rush town, but using the world of boxing to bring even more attention to the town and its mines would result in riches for all.

The fighter invited to defend his title was Joe Gans, a Baltimore native born in 1874. Growing up, he had been a part of "Battle Royales," where 15 or so black teenage boys were blindfolded and thrown into a ring for a slugfest, with the last one standing declared the winner. The white male spectators would bet on which of the young men would survive the brutal beating. But Gans emerged from those days wanting to learn boxing skills and sought out a mentor. Early in his life, he also took on a manager who contracted his bookings from his days as a teenager until Gans fired him just before he was invited to Goldfield. In 1902, he had won the lightweight title, and by 1906, at 32, had fought 145 times, only losing six bouts. Now Gans would defend his title against Oscar Nelson, who was known as a dirty fighter and virulent racist.

George Lewis "Tex" Rickard was born in Kansas City, Missouri, on January 2, 1870, on a ranch next to the homestead of the parents of famed outlaw Jesse James. Rickard's life seemed to embody every bit of the Wild West, from riding in a covered wagon with his family to Texas, to becoming a cowboy, to prospecting in the Klondike Gold Rush. When he arrived in Goldfield, he set up the Northern saloon, and his true potential began to emerge. But to achieve success, he needed the cooperation of the reigning lightweight champ. (LOC.)

Joe Gans was born on November 25, 1874, in Baltimore, Maryland. His first job was at a fish market where the owner mentored him in boxing. Gans, like many boxers of the day, boxed in theaters for entertainment. Gans was a technician. He studied other boxers and delivered a straight knockout punch. He is said to have developed the upper cut as a consistent technique and was one of the first to use the clinch to effect. His studious nature and gentlemanly conduct earned him the name "The Old Master." He never weighed over 137 pounds, a lightweight. In July 1906, Rickard found Gans in San Francisco and invited him to fight in Goldfield. (LOC.)

Oscar Mattheus Nielsen was born on June 5, 1882, in Denmark and got the nickname "Battling" from his dad who said he was born battling. After moving to Chicago in 1883, the last name was changed to Nelson, and as a kid, Oscar earned money hauling ice. In 1896, he got in the ring with a strong man at a circus and knocked him out. He started taking other fights that lasted from 16 to 20 rounds, earning him the title "The Durable Dane." He regarded all black fighters as inferior and was a hero to the Ku Klux Klan. In 1906, Rickard secured Nelson to fight Gans by making a deal with his manager, Billy Nolan. Nelson arrived in Goldfield on August 11, 1906, in this Pope Toledo and said he "didn't care to travel any faster in a chug-chug wagon." Billy Nolan is behind Nelson holding his hat, and Ole Elliott is just above Nolan. Right of Elliott is Tex Rickard, and Charles Chrisman is driving. At the corner of Main and Crook Street (Highway 95 today), the Palace is in the background. (CNHS.)

The group that put on the contest, the Goldfield Athletic Club, is assembled here for the signing of the articles of the fight. Oscar Nelson (seated left) and Joe Gans (seated right) review the document as Tex Rickard sits left of Gans and Nelson's manager Billy Nolan is left of Rickard. Standing right of Gans is Ole Elliott, who had one of Goldfield's first saloons and was one of the town's founding members. Standing behind Gans is Larry Sullivan, who appointed himself Gans's manager while in Goldfield. (SCP.)

SYNOPSIS

Contestants—
  "Battling" Nelson
  "Joe" Gans

Purse—$30,000

Nelson's Share—
  $20,000, win or lose

Gans' Share—
  $10,000 win or lose

Weight at Ringside—
  13ᵒ pounds

Length of Contest—
  To a finish

Referee—
  George Siler

Master of Ceremonies—
  L. M. Sullivan

Calendar of Events Preliminary to Contest

July 30—Goldfield Athletic Club organized.

July 30—"Tex" Rickard wires for fight between Britt and McGovern.

July 31—$5,000 purse guaranteed by wire for fight.

Aug. 1—Nolan wired Club that Nelson would fight Clifford.

Aug. 2—$20,000 offered by Club for Nelson-Gans match.

Aug. 3—Nolan wires demand for $30,000 purse.

Aug. 3—$52,000 raised among Athletic Club members in three hours; a total of $110,000 was raised in twenty-four hours.

Aug. 4—Gans wired he would meet Nelson. Thirty thousand dollars for purse deposited in John S. Cook Bank.

Aug. 6—San Francisco offered $40,000 and Sacramento $42,500 for the match. Nolan wires he is en route to Goldfield.

Aug. 7—"Tex" Rickard meets Nolan at Reno. Gans arrived in Goldfield.

Aug. 8—Rickard and Nolan arrived.

Aug. 9—Gans and Nolan disagree over purse. Gans designates L. M. Sullivan his manager.

Aug. 10—Preliminary agreement as to purse signed by Nolan and Gans.

Aug. 11—Nelson arrived. Articles of agreement signed.

NOTE—This, the greatest glove contest for the lightweight championship ever held in America, was arranged and "signed up" in the remarkably short period of twelve days. The purse is the largest guaranteed sum ever offered for such an event.

The terms of the fight as well as the timeline are laid out explicitly in this page from the official fight program. The purse was the largest in prize fight history at the time. Nelson received $20,000 win or lose as negotiated between Rickard and Nelson's manager Nolan. Gans's share was $10,000 win or lose. (UNLV.)

This rare portrait of Joe Gans shows his undying positive outlook while in Goldfield. In fact, one reporter described him as "the happiest man in camp." Gans arrived in Goldfield on August 7, 1906, several days before Nelson. The Tonopah & Goldfield train pulled into town at 9:15 p.m., most likely at the station across from the Santa Fe Saloon. Author William Gildea described a "crowd of miners gathered at the depot—grizzled white men who had drifted to the edge of civilization and lived in tents or hollowed-out hills . . . They looked tough enough to dig all day or fight one another for an ounce of gold—men of Battling Nelson's stripe. And yet they cheered Gans, jostling to get a look at the newest visitor . . . They understood him to be a gentleman in the ring who observed rules that others such as Nelson routinely ignored . . . and they were ready to root for him." (SCP.)

Larry Sullivan became Gans's manager as Gans was required to put up $5,000 to guarantee he would show up for the fight. As Gans did not have the money, Sullivan paid it in exchange for managing him. William Gildea quotes Sullivan: "If you lose, you'll never get out of Goldfield alive. My friends are going to bet a ton of money on you. They will kill you if you don't beat Nelson." Gans took the threat with a laugh. "If I had any money," Gans replied calmly, "I'd bet it on myself." (CNHS.)

Merchants Hotel

E. J. SCHMIDT, Proprietor

Telephone 642

One Block From Goldfield &
Columbia R. R. Depot

Automobile and Carriages
Drive to and from Goldfield
to Columbia.

First Class Hotel
Rates Reasonable

Rickard arranged for Gans to stay at the Merchants Hotel in nearby Columbia, about a mile from Goldfield. William Gildea states, "Gans stayed in a three-room suite, which was described as comfortable and quiet. A gym was built behind the hotel for him, and he soon got into a routine of taking a nine-mile run back and forth at 7:00 every morning, then came back to the hotel for a bath, a rubdown, and lunch. Visitors came to the hotel and watched him work out with the punching bag or skipping rope in repetition more than 1,000 times. Visitors found Gans extremely cordial, and he was even visited by Nevada governor John Sparks." (CNHS.)

The interior of the hotel featured gaming tables and an upright piano. While Gans did not fare well at gambling, he enjoyed the sounds of the piano and sometimes danced and sang to the music. Gans's female companion, Martha Davis, was an accomplished musician who played the piano in the lobby for the entertainment of the guests. The press and those who met her referred to her as Mrs. Gans, though she was not. Joe was in the midst of a divorce from his wife, Madge Watkins, back in Baltimore, and Martha accompanied him to his fights in the West and was ringside in Goldfield. (CNHS.)

106

Gans sparred and trained extensively in the month prior to the fight. He had to make the 133-pound weight limit, and only he, not Nelson, would be weighed three times on the day of the contest, including just minutes before the fight. The last weigh-in was intended to weaken Gans, as he would have no time to rehydrate sufficiently. On top of that, Nolan insisted Gans weigh in wearing trunks, gloves, and shoes, meaning his actual weight had to be lower than 133. Gans responded, "I'll weigh in harness. I'll agree to anything to have the fight come off." Those arrangements were made by Nolan and agreed to by Sullivan for Gans. (SC.)

Nelson had trouble finding headquarters but settled on the lower floor of the Ladies Aid Hall. He was eight years younger than Gans and reveled in long fights. Sally Zanjani remarks, "one of his fights, in which he and his opponent had been floored forty-nine times, held the alltime record for knockdowns." William Gildea quotes Nelson saying, "I ain't human," and parts of the medical community agreed. A blow to the head that might render other men unconscious had no effect on Nelson. (SCP.)

Larry Sullivan (left) and Tex Rickard (right) are seen here with Oscar Nelson behind a stack of 1,500 twenty-dollar coins representing the $30,000 purse. About 10 days prior to this photograph, on August 4, 1906, the *Goldfield Sun* declared "The Battle of the Century! Battling Nelson and Joe Gans will engage in a finish fight on Labor Day . . . the purse of $30,000 is in the Cook bank. Tex Rickard declared himself in for ten thousand dollars for the purse. Seven prominent businessmen subscribed twenty thousand dollars in a few minutes." (CNHS.)

While Nelson was doing promotional shots at the bank, Gans was taking in the scenery. Both fighters participated in photo opportunities such as helping to clean a street or hoisting an ore bucket, but Gans enjoyed getting to know the town. Here he is seen with a Goldfield local. (TEC.)

This image is labeled "Allen Photo" (Arthur Eppler) and dated July 4, 1906, but there is no record that Gans was in Goldfield on that date. He must have made a quick trip to Goldfield, as he is recorded as being in San Francisco around July 1. His first documented arrival in town is August 7, 1906, but here he is in an REO Model T with Charles Chrisman and a fight organizer standing at right. (UNLV.)

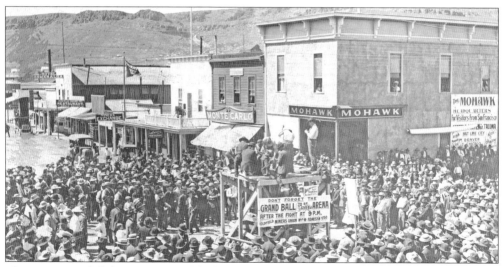

The *Sun* reported: "The streets of Goldfield were jammed . . . Drilling contests preceded the battle. Waters and Hill of Tonopah drilled thirty-seven inches in fifteen minutes . . . followed by burro races and foot-races." In drilling contests, men competed to see how deep they could drill into a rock using hammers and a pointed steel rod. The view is Crook and Main Streets. (TEC.)

Real Money of $30,000 Is Up and There Is No String to It---No Reason to Believe That the Contest Is Not on the Level

Nelson punching the bag

B ATTLING NELSON and Joe Gans are to battle for the lightweight championship of the world at Goldfield, Nevada, tomorrow afternoon. They will battle for the biggest purse ever offered little men—$30,000. $20,000 of which goes to Nelson, win or lose. Bob Fitzsimmons and Jim Hall were supposed to have battled for a $40,000 purse at New Orleans, but Fitz, the winner, never saw the money. In the present instance the money —real money—is up. There is no string to it.

GREAT INTEREST IN THE FIGHT

Outside of a fight in which James J. Jeffries would figure, it is doubtful if a bout could be arranged which could arouse the interest that this one has. There is something about the rugged Dane that has won for him the admiration of the sport-loving public. He doesn't claim anything but a fighter. He has gone right down the line meeting and beating all comers, and tomorrow he is to meet one of the greatest pugilists that ever pulled on a glove—Joe Gans.

There never was a question about the ability of Gans—when he was right. Since Kid Lavigne, through his own dissipation, passed up the lightweight crown, there never has been a lightweight in Gans' class. H was so good, in fact, that he had to go outside of his class to get on a match, and even then he was compelled to agree to frame-ups, or go without the coin. He was so good that he had to work himself out of the chance to make money on the level. He lay down to Terry McGovern at Chicago, and he pulled to Jimmy Britt in 'Frisco. Before that he lay down to Frank Erne at the Broadway Club in New York, and a year later at Fort Erie he stopped him in a round.

CONFESSION GOOD FOR THE SOUL

But Gans has made the honest confession which is said to be good for the soul. Hereafter he does not propose to listen to the tempters—the sure thing players, who have profited by his crookedness in the past. He is going to stick to the straight and narrow path. He will have to be taken at his word. Ac-

---

HEAR ABOUT THE FIGHT
AT THE INQUIRER OFFICE

Full Returns of the Fistic Battle between Joe Gans and Battling Nelson, at Goldfield, Nev., will be megaphoned from the front of

The Inquirer Office, Monday Evening

The men enter the ring at **3 P. M.**, which means that the first returns will be received in this city about **6 P. M.**, owing to the difference in time.

All are invited to come around and hear not only the fullest and most complete account of each round, but also to hear it at the same time that it

Happens in the Ring.

The Inquirer is always first and foremost, and you will hear about the fight first at

——THE INQUIRER OFFICE——

---

cording to the articles of agreement and the conditions which govern his bout tomorrow afternoon, it would be simply suicidal for him to do anything but put up the best he has in the shop. Win or lose, he can only get McCormick's end of the purse—$10,000. There is every inducement for him to win if it is within his ability to do so. A positive victory over Nelson would be worth more to him than all the sure thing players could possibly offer him to throw the fight. So far as Nelson is concerned, a victory over Gans would make him the undisputed lightweight champion, and a title worth many times more money than any jobbers could possibly afford to offer him to throw the fight, even if he and his manager, Billy Nolan, were willing to consider any proposition. There isn't the slightest reason for anticipating anything but an up-and-up fight from beginning to end.

EXPERT OPINION FAVORS GANS

With a fair field and no favor, the question is, who will be the victor?

This is the *Philadelphia Inquirer* from Sunday morning, September 2, 1906, the day before the fight. Interest in the fight had gone nationwide. Note the insert, "Hear About The Fight." Full returns would be megaphoned from the front of the paper's office. It was possible to hear the fight blow-by-blow being announced as it happened. This scene was repeated in major cities across the country.

Bal in action.

Gans in action.

Battling Nelson.

Joe Gans.

The paper extended an invitation: "all are invited to come around and hear not only the fullest and most complete account of each round, but also to hear it at the same time that it *Happens in the Ring*." It was the first time a boxing match was carried live across the country. (LOC.)

There was some controversy over who the referee would be, but finally the Goldfield Athletic Club decided on George Siler, seen here between the boxers. The man behind Nelson with the wavy hair is Larry M. Sullivan, who stepped down as president of the club when he became Gans's manager. Sullivan, described as a "loudmouth," wanted to emcee the event, which he did. He read telegrams from well-wishers, one of whom was Gans's mother: "Joe, the eyes of the world are on you . . . you bring back the bacon," which is where the saying began. The event was filmed by Miles Brothers, and Percy Dana was the official photographer for the fight and program. It was a hot September day when the fight got underway around 3:20 p.m. Gans's trunks were blue, and Nelson's were green, red, white, and blue. Nelson fought his dirtiest, and Gans fought his best. The crowd booed and jeered Nelson's dirty tricks and cheered Gans's skill and tactics. Elvira Nelson was a little girl who snuck into the fight. In 1988, she recalled, "The fight lasted for hours. It was hot and we tried to leave but the gate was locked and we had to wait until the fight was finished to get out. I remember all the blood." The fight lasted 42 rounds—the longest in boxing history. (TEC.)

Gans was hurt in the 33rd round when he broke his right hand, but pretended to have a sprained foot instead so as not to show weakness. Nelson fouled him with a low blow in the 42nd round, and Gans won the fight. The event was a success, with total gate receipts of $72,000, the highest on record. Larry Sullivan's company received a windfall, as his business partner George Graham Rice wagered $45,000 on Gans. (TEC.)

Joe brought home the bacon, and became America's first black boxing champion. He went back to Baltimore and built the Goldfield Hotel in honor of his time in Nevada. As a "black-and-tan" hotel, it catered to mixed crowds, and famed jazz musician Eubie Blake got his start there. But Gans was suffering from tuberculosis. It is likely he had it in Goldfield yet continued to take on challengers, fighting several more years. When he died in 1910, his funeral was fit for a king with hundreds of carriages lining the streets, and more than 5,000 people attending. In 1990, he was inducted into the International Boxing Hall of Fame. (LOC.)

Six

DOOMED

The story of America's western mining camps and towns is typically a tale of discovery, boom, then bust. Goldfield was no different in terms of the cycle. It was significantly different in terms of its size and scope. There is no doubt that 1906 was the year when everything came together for Goldfield. The Mohawk Mine had made the richest discovery to date. The Gans-Nelson fight had earned the city national attention. Just two months after the fight, George Wingfield and Sen. George Nixon created the Goldfield Consolidated Mining Company, one of the largest mining corporations in the United States. From 1906 into most of 1907, Goldfield seemed unstoppable. Business was booming, and buildings were being constructed . . . with a combustible rock called white rhyolite.

Fires plagued every mining town. After all, they were mostly made of wood, kerosene lamps and candles were everywhere, and the new autos were running on gasoline. What could go wrong? Goldfield's first fires came in 1905. One was confined to a bank, another consumed two business blocks, and the third was in Columbia, where fire tore down Main Street while 200 men in a bucket brigade saved the rest of the town from disaster, including the Merchants Hotel where boxer Joe Gans would stay the following year. Shortly after the Gans-Nelson fight, Goldfield was hit by two more fires.

And then there was the flood. A sudden cloudburst caused a steady downpour on September 13, 1913. Goldfield, being in a basin, acted as a funnel for the water coming from the mountains. While the business districts and Main Street were on higher ground, the flood took out rail lines and homes and businesses situated in lower parts of the basin.

By the 1920s, the mines were played out, the economy was in a slump, fires and flood had taken their toll, and the population had moved on, but some of the town's most elegant stone buildings still stood. And then, it happened. Author Hugh Shamberger describes it as "one of the most destructive fires ever to occur in a Nevada mining town." That is saying something, as Nevada had its fair share of dramatic mining town conflagrations. The fire of 1923 started across the street from the Goldfield Hotel and quickly took out more than 200 structures. The white rhyolite used in the stone structures made the buildings literally explode. That fire was followed up by another in 1924, which destroyed the elegant News Building. Goldfield, it seemed, was doomed.

ALLEN·PHOTO
CO -

DESTRUCTION -OF·THE·
- NOV·17·1906

DFIELD-HOTEL
-DFIELD-NEV -

This Arthur Eppler photograph shows the destruction of the Goldfield Hotel one year after its opening on November 1, 1905. Two died in this blaze, with losses at $50,000 ($1.5 million today). The fire was limited to the hotel. (SCP.)

The 1913 flood took out homes and damaged railroad tracks. Fortunately, it left most of the businesses in the main part of Goldfield intact, as they were on higher ground. (SC.)

The final nail in Goldfield's coffin came in 1923 and 1924, when two great fires destroyed much of what was left of the old town. At far right is the Goldfield Hotel. The brick hotel, built on the site of the old Goldfield Hotel that burned in 1906, escaped without any significant damage. The hollowed-out building across the street was the News Building, whose magnificent arches and stone structure could not withstand the fire's heat. The Ish-Curtis Building to the left sustained damage to the roof and third floor but was determined sound enough that it could be repaired. But for the most part, Goldfield's old business district was gone. (TEC.)

Seven

REVIVAL

It is always important to note that there was a population of Native Americans who knew and understood this remote desert for many hundreds, if not thousands, of years before Goldfield came along. But from the earliest prospecting venture near Columbia Mountain until today, the town known as Goldfield has always had some population. At its peak in 1907 or so, there were around 20,000, yet today only a few hundred call Goldfield home, and most have the same grit and resilience as the early Goldfielders. Yet Goldfield always had investors from distant cities who took an interest in the town. The same is true today.

There are many committed to reviving Goldfield. There are some who can trace their roots to the gold rush days of 1905–1907, like Allen Metscher, whose family has likely done more to preserve the history of Goldfield and Tonopah than anyone. They are responsible for the Central Nevada Museum and Central Nevada Historical Society in Tonopah. There are others who have come since then and stayed, others who may live in Goldfield and have another home, or some who just live elsewhere and want to help restore some of Goldfield to its original splendor. There are many committed to reviving Goldfield. The properties and artifacts like the traction engine featured in this chapter go back to the boom days of 1904–1910.

Some, like Carl Brownfield, keep Goldfield in the media with his radio station KGFN, and the Mozart, run by Gina Greenway, keeps the food flowing. The Dinky Diner also serves up good Goldfield fare as well. And Steve and Jeri Foutz, who recently moved to Goldfield, are restoring a 1907 boardinghouse and have created a small resort with cabin rentals and an RV park.

Many who share the goal of reviving Goldfield are from Los Angeles, Las Vegas, Sacramento, Salt Lake City, or other far-flung cities, but they share a common goal. They've purchased the remaining intact buildings and are committed to restoring them. It is an unusual passion to want to breathe life into a ghost town, and everyone who does it seems to thrive off of seeing history come to life in real and tangible ways.

Those restoring and telling the stories of these places and artifacts are featured here as part of Goldfield's ongoing revival.

Kim Aurich is on the porch of one of Goldfield's original homes, being restored by the Aurich family. Built between 1906 and 1908, it belonged to Tom Lockhart, one of the first to arrive in Goldfield in 1903. He sold his mining property in Tonopah and invested in the mines called the Florence Group. In 1905, the Florence-Goldfield Mining Company was formed with Lockhart, George Wingfield, Sen. George Nixon, and other investors. (TF.)

James Aurich and his father, Jon, are seen here with the workings of Lockhart's Florence Mine in the background. The Aurich family is dedicated to both preserving and telling the extraordinary tale of Goldfield. The family maintains homes in both Las Vegas and Goldfield and preserves property and artifacts and offers tours of the mine, led by James. (TF.)

The Santa Fe Saloon, established in 1905, is Nevada's second-oldest continually operated saloon. Its owner, Jim Marsh, a Las Vegas resident, takes great pride in procuring and protecting some of Nevada's historic properties. From southern to central and northern Nevada, Marsh maintains many historic sites for the enjoyment and use of future generations. (TF.)

Tex Rickard's brick house, built in 1906, has stained-glass windows and is elegant for any era. It does not, however, have a kitchen. Since Rickard owned a restaurant, he likely did not see a kitchen as essential. Rickard went on to become the country's premier boxing promoter, taking on Reno's famed Johnson-Jeffries fight of 1910 as well as other well-known bouts. He went to New York, where he built the third incarnation of Madison Square Garden, which stood from the 1920s until 1968. Rickard died on January 6, 1929, at the age of 59 due to complications from an appendectomy. The current owner of his Goldfield home is restoring it to secure its future. (TF.)

Don Woolcott stands outside the original Goldfield jail, which he now owns. Woolcott lives in Northern California but often visits to work on the jail. (TF.)

Glenn Praiss owns Goldfield's Consolidated Telephone-Telegraph Company Building, which operated from 1908 to 1963. It was in one of the blocks spared by the 1923 fire, and as the sign states, "It is an unspoiled expression of the work of turn of the century craftsmen, and serves as an example of the business life in the area." Glenn owns a number of historic properties in town that he lovingly restores and preserves. (TF.)

Rich and Edie Koepnick bought one of the brothels in Goldfield's red-light district. Their goal is to present the story of the "ladies of the night" who worked in that part of town. Though Rich has passed away since this photograph was taken, Edie still maintains her connection to the town and this historic building. (TF.)

Steve Roberts came from Utah and fell in love with Goldfield and its story. He purchased the historic Downer Bros. Assay Office and has restored it to its former glory. (TF.)

Randy Main built the Palace to hold his collection of period artifacts and to have a place to stay when in Goldfield. He entertains people by sharing his collection and also lets the place be used for local events. Here, he stands in front of one of his most iconic artifacts, a steam traction engine used to haul goods to and from Goldfield during the boom days. (TF.)

This image features a steam traction engine in action hauling goods to Goldfield around 1907. The Best Company manufactured many of these engines in that era; today, Caterpillar Inc. is its descendant. (CNHS.)

The historic Ish-Curtis Building (constructed 1907), which in 1919 came to house the John S. Cook Bank, is owned by Telly and Caroline Eliades. They enjoy collecting all things Goldfield, from photographs to artifacts to its historic buildings. The couple is committed to keeping history alive by not only preserving history, but making their collections and buildings available to the public. Caroline recreates historic clothing (often from only a photograph and without a pattern). This building is now home to a magnificent Brunswick bar, which had belonged to Wyatt Earp at his Tonopah bar, the Northern, then travelled to Rhyolite and on to Las Vegas and a few other places before being purchased by Telly and Caroline and moved to Goldfield. (TF.)

John Ekman, seen in front of Goldfield High School, is the current president of the Goldfield Historical Society.

In what can only be described as a heroic effort, he has undertaken the task of restoring Goldfield High School. Though nearly an entire wall was missing, Ekman saw the remarkable beauty of a well-preserved interior and potentially restored exterior.

With help from the community, along with local and state offices and others, the historic high school is being saved for future generations. (TF.)

John Ekman is seen here in one of the school's well-preserved classrooms. While he and all involved with this project acknowledge the work is not yet done, Ekman retains an enthusiasm for reviving and restoring Goldfield's gems. From a historic rail yard to some remarkable homes, Ekman's impact on preserving Goldfield is well respected by the local community and the state of Nevada. He, along with many others, are the people who are helping bring about Goldfield's revival. (TF)

BIBLIOGRAPHY

Aycock, Colleen, and Mark Scott. *Joe Gans: A Biography of the First African American World Boxing Champion.* Jefferson, NC: McFarland and Company Inc., 2008.

Crampton, Frank. *Deep Enough.* Norman, OK: University of Oklahoma Press, 1993.

Earl, Phillip I. *This Was Nevada.* Reno, NV: Nevada Historical Society, 1986.

Elliott, Russell R. *Nevada's Twentieth-Century Mining Boom.* Reno, NV: University of Nevada Press, 1966.

Fleischer, Nat. *The Three Colored Aces: Story of George Dixon, Joe Gans, and Joe Walcott and Several Contemporaries.* New York, NY: The Ring Publishing, 1938.

Gildea, William. *The Longest Fight: In the Ring with Joe Gans, Boxing's First African American Champion.* New York, NY: Farrar, Straus, and Giroux, 2012.

Glasscock, C.B. *Gold in Them Hills.* New York and Chicago: A.L. Burt Co., 1932.

————. *Here's Death Valley.* Indianapolis and New York: Bobbs-Merrill Co., 1940.

www.goldcreekfilms.com

www.goldfieldhistoricalsociety.com

Goldfield Historical Society. *Goldfield Historic Walking Tour Booklet.* Goldfield, NV: Goldfield Historical Society, 2013.

————. *Goldfield Remembered.* Goldfield, NV: Goldfield Historical Society, 1994.

James, Ronald M. *Temples of Justice: County Courthouses of Nevada.* Reno, NV: University of Nevada Press, 1994.

Lingenfelter, Richard E. *Death Valley and the Amargosa: A Land of Illusion.* Berkeley and Los Angeles: University of California Press, 1986.

Patera, Alan H. *Goldfield's Fabulous Boom.* Lake Oswego, OR: Western Places, 2013.

Provost, Stephen H. *Goldfield Century.* Fresno, CA: Century Cities Publishing, 2021.

Raymond, C. Elizabeth. *George Wingfield: Owner and Operator of Nevada.* Reno, NV: University of Nevada Press, 1992.

Rickard, Maxine Elliott Hodges. *Everything Happened to Him: The Story of Tex Rickard.* New York, NY: Frederick A. Stokes Company, 1936.

Samuels, Charles. *The Magnificent Rube: The Life and Gaudy Times of Tex Rickard.* New York, NY: McGraw-Hill, 1957.

Shamberger, Hugh A. *Goldfield.* Carson City, NV: Western Printing and Publishing, 1982.

Thornton, T.D. *My Adventures With Your Money: George Graham Rice and the Golden Age of Con Artistry.* New York, NY: St. Martin's Press, 2015.

Zanjani, Sally. *Another Life: Tales of Nevada's Last Gold Rush.* Reno, NV: Nevada Publications, 2018.

————. *The Glory Days of Goldfield Nevada.* Reno, NV: University of Nevada Press, 2002.

————. *Goldfield: The Last Great Gold Rush on the Western Frontier.* Athens, OH: Swallow Press/ Ohio University Press, 1992.

DISCOVER THOUSANDS OF LOCAL HISTORY BOOKS FEATURING MILLIONS OF VINTAGE IMAGES

Arcadia Publishing, the leading local history publisher in the United States, is committed to making history accessible and meaningful through publishing books that celebrate and preserve the heritage of America's people and places.

Find more books like this at
www.arcadiapublishing.com

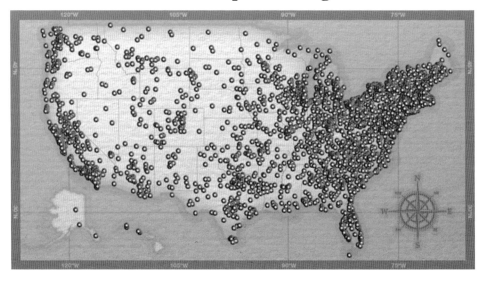

Search for your hometown history, your old stomping grounds, and even your favorite sports team.

Consistent with our mission to preserve history on a local level, this book was printed in South Carolina on American-made paper and manufactured entirely in the United States. Products carrying the accredited Forest Stewardship Council (FSC) label are printed on 100 percent FSC-certified paper.

MADE IN THE USA